Official

SQA Past Papers WITH ANSWERS

Higher
Chemistry

2010–2014

HODDER
GIBSON
AN HACHETTE UK COMPANY

Hodder Gibson is grateful to the copyright holders, as credited on the final page of the Question Section, for permission to use their material. Every effort has been made to trace the copyright holders and to obtain their permission for the use of copyright material. Hodder Gibson will be happy to receive information allowing us to rectify any error or omission in future editions.

Hachette UK's policy is to use papers that are natural, renewable and recyclable products and made from wood grown in sustainable forests. The logging and manufacturing processes are expected to conform to the environmental regulations of the country of origin.

Orders: please contact Bookpoint Ltd, 130 Park Drive, Abingdon, Oxon OX14 4SE. Telephone: (44) 01235 827720. Fax: (44) 01235 400454.

Lines are open 9.00–5.00, Monday to Saturday, with a 24-hour message answering service. Visit our website at www.hoddereducation.co.uk. Hodder Gibson can be contacted direct on: Tel: 0141 848 1609; Fax: 0141 889 6315; email: hoddergibson@hodder.co.uk

This collection first published in 2014 by

Hodder Gibson, an imprint of Hodder Education,

An Hachette UK Company

2a Christie Street

Paisley PA1 1NB

{BrightRED   Hodder Gibson is grateful to Bright Red Publishing Ltd for collaborative work in preparation of this book and all SQA Past Paper, National 5 and Higher for CfE Model Paper titles 2014.

Typeset by PDQ Digital Media Solutions Ltd, Bungay, Suffolk NR35 1BY

Printed in the UK

A catalogue record for this title is available from the British Library

ISBN 978-1-4718-3678-7

3 2 1

2015 2014

Introduction

Study Skills – what you need to know to pass exams!

Pause for thought

Many students might skip quickly through a page like this. After all, we all know how to revise. Do you really though?

Think about this:

"IF YOU ALWAYS DO WHAT YOU ALWAYS DO, YOU WILL ALWAYS GET WHAT YOU HAVE ALWAYS GOT."

Do you like the grades you get? Do you want to do better? If you get full marks in your assessment, then that's great! Change nothing! This section is just to help you get that little bit better than you already are.

There are two main parts to the advice on offer here. The first part highlights fairly obvious things but which are also very important. The second part makes suggestions about revision that you might not have thought about but which WILL help you.

Part 1

DOH! It's so obvious but …

Start revising in good time

Don't leave it until the last minute – this will make you panic.

Make a revision timetable that sets out work time AND play time.

Sleep and eat!

Obvious really, and very helpful. Avoid arguments or stressful things too – even games that wind you up. You need to be fit, awake and focused!

Know your place!

Make sure you know exactly **WHEN and WHERE** your exams are.

Know your enemy!

Make sure you know what to expect in the exam.

How is the paper structured?

How much time is there for each question?

What types of question are involved?

Which topics seem to come up time and time again?

Which topics are your strongest and which are your weakest?

Are all topics compulsory or are there choices?

Learn by DOING!

There is no substitute for past papers and practice papers – they are simply essential! Tackling this collection of papers and answers is exactly the right thing to be doing as your exams approach.

Part 2

People learn in different ways. Some like low light, some bright. Some like early morning, some like evening / night. Some prefer warm, some prefer cold. But everyone uses their BRAIN and the brain works when it is active. Passive learning – sitting gazing at notes – is the most INEFFICIENT way to learn anything. Below you will find tips and ideas for making your revision more effective and maybe even more enjoyable. What follows gets your brain active, and active learning works!

Activity 1 – Stop and review

Step 1

When you have done no more than 5 minutes of revision reading STOP!

Step 2

Write a heading in your own words which sums up the topic you have been revising.

Step 3

Write a summary of what you have revised in no more than two sentences. Don't fool yourself by saying, "I know it, but I cannot put it into words". That just means you don't know it well enough. If you cannot write your summary, revise that section again, knowing that you must write a summary at the end of it. Many of you will have notebooks full of blue/black ink writing. Many of the pages will not be especially attractive or memorable so try to liven them up a bit with colour as you are reviewing and rewriting. **This is a great memory aid, and memory is the most important thing.**

Activity 2 — Use technology!

Why should everything be written down? Have you thought about "mental" maps, diagrams, cartoons and colour to help you learn? And rather than write down notes, why not record your revision material?

What about having a text message revision session with friends? Keep in touch with them to find out how and what they are revising and share ideas and questions.

Why not make a video diary where you tell the camera what you are doing, what you think you have learned and what you still have to do? No one has to see or hear it, but the process of having to organise your thoughts in a formal way to explain something is a very important learning practice.

Be sure to make use of electronic files. You could begin to summarise your class notes. Your typing might be slow, but it will get faster and the typed notes will be easier to read than the scribbles in your class notes. Try to add different fonts and colours to make your work stand out. You can easily Google relevant pictures, cartoons and diagrams which you can copy and paste to make your work more attractive and **MEMORABLE**.

Activity 3 – This is it. Do this and you will know lots!

Step 1

In this task you must be very honest with yourself! Find the SQA syllabus for your subject (www.sqa.org.uk). Look at how it is broken down into main topics called MANDATORY knowledge. That means stuff you MUST know.

Step 2

BEFORE you do ANY revision on this topic, write a list of everything that you already know about the subject. It might be quite a long list but you only need to write it once. It shows you all the information that is already in your long-term memory so you know what parts you do not need to revise!

Step 3

Pick a chapter or section from your book or revision notes. Choose a fairly large section or a whole chapter to get the most out of this activity.

With a buddy, use Skype, Facetime, Twitter or any other communication you have, to play the game "If this is the answer, what is the question?". For example, if you are revising Geography and the answer you provide is "meander", your buddy would have to make up a question like "What is the word that describes a feature of a river where it flows slowly and bends often from side to side?".

Make up 10 "answers" based on the content of the chapter or section you are using. Give this to your buddy to solve while you solve theirs.

Step 4

Construct a wordsearch of at least 10 X 10 squares. You can make it as big as you like but keep it realistic. Work together with a group of friends. Many apps allow you to make wordsearch puzzles online. The words and phrases can go in any direction and phrases can be split. Your puzzle must only contain facts linked to the topic you are revising. Your task is to find 10 bits of information to hide in your puzzle, but you must not repeat information that you used in Step 3. DO NOT show where the words are. Fill up empty squares with random letters. Remember to keep a note of where your answers are hidden but do not show your friends. When you have a complete puzzle, exchange it with a friend to solve each other's puzzle.

Step 5

Now make up 10 questions (not "answers" this time) based on the same chapter used in the previous two tasks. Again, you must find NEW information that you have not yet used. Now it's getting hard to find that new information! Again, give your questions to a friend to answer.

Step 6

As you have been doing the puzzles, your brain has been actively searching for new information. Now write a NEW LIST that contains only the new information you have discovered when doing the puzzles. Your new list is the one to look at repeatedly for short bursts over the next few days. Try to remember more and more of it without looking at it. After a few days, you should be able to add words from your second list to your first list as you increase the information in your long-term memory.

FINALLY! Be inspired...

Make a list of different revision ideas and beside each one write **THINGS I HAVE** tried, **THINGS I WILL** try and **THINGS I MIGHT** try. Don't be scared of trying something new.

And remember – "FAIL TO PREPARE AND PREPARE TO FAIL!"

Higher Chemistry

The exam

You have worked hard in class and at home, and the Higher Chemistry exam will give you the opportunity to show what you know. The examiners want candidates to do as well as possible and while there may be some tricky questions, there are never any trick questions. This section will help you to make sure that you are well prepared for the exam and can receive full credit for all of the hard work you have already done.

Being prepared…

The Higher Chemistry exam follows the same format each year and, by using this past paper booklet, you can become really familiar with what to expect.

Section A

Section A is the multiple choice section. Near the beginning of the paper there are often a couple of questions testing ideas you met as part of the Standard Grade or Intermediate 2 course. Don't be too worried if, amongst the first few questions, you see something that you haven't studied this year. An example of this would be question 3 from the 2013 paper.

3. Which of the following would be expected to react?

    A   Iron and zinc sulphate solution

    B   Tin and silver nitrate solution

    C   Copper and dilute sulphuric acid

    D   Lead and magnesium chloride solution

This question is testing your knowledge of displacement reactions from Standard Grade/Intermediate 2.

Where you can achieve extra marks: when you are confident that you have revised all of the Higher content to the best of your ability, brush up on the areas of the Standard Grade and Intermediate 2 courses which commonly appear in this part of the paper. Areas to revise would include; corrosion, displacement reactions, precipitation and carbohydrates.

Section A usually follows the order of material as presented in the course units. By looking through this past paper booklet you will get a good feel for the types of questions you are likely to encounter.

Section B

Section B is the "written paper". Each Higher paper will generally tend to contain a number of different types of questions designed to test different skills. While you will be able to answer some questions simply by remembering a fact or a name, some might ask for a more detailed explanation and some will give you the chance to show that you can solve problems.

Calculations

Some people are surprised to discover that around twenty marks in the Higher Chemistry exam are awarded for calculations. It is an extremely good idea to get as much practice tackling this type of question as possible. Higher Chemistry papers will usually include questions testing certain standard types of calculations. These would normally include:

- a calculation involving either relative or average rate e.g. 2013, 2(b)(i)
- a calculation involving Avagadro's number (6.02×10^{23}) e.g. 2013, 3(b)
- a calculation involving the measurement of an enthalpy change e.g. 2013, 15(a)
- a calculation involving the molar volume of gases e.g. 2013, 13(b)
- a percentage yield calculation e.g. 2013, 11(b)
- a calculation based on half-life e.g. 2013, 12(b)
- a Hess's law calculation e.g. 2013, 14(a)(ii)
- a calculation based on an electrolysis experiment e.g. 2013, 17(a)

Markers will be looking for reasons to award marks, so even if you don't get the correct final answer you will get credit for any working that you have written that shows you are on the right track. Markers will also award marks for "follow through". This means that even if something goes wrong early in a calculation, the markers will work with whatever values you have come up with in the later stages of the calculation to try to award marks.

Where you can achieve extra marks: even if you are confident that you can do a certain type of question, always show all of your working in your answers. This way, even if you make mistakes, the marker will still be able to award partial marks for any correct working shown.

Prescribed Practical Activities

Every Higher student should have completed nine Prescribed Practical Activities as part of the course. These are described both in textbooks and in revision guides. In each paper, knowledge of at least three of the PPA experiments will usually be assessed. Some questions clearly name the PPA being described, such as Question 2, 2013. In others, you need to show that you can apply your knowledge of the PPA in another situation, for example Question 16(b)(i), 2013.

Where you can achieve extra marks: with around six marks in each Higher paper awarded for knowledge of the Prescribed Practical Activities, you should make sure that you refresh your memory of the PPA experiments you performed as part of your course.

Problem Solving within the context of experimental design

There is usually a problem-solving question that requires candidates to think about an unfamiliar experiment, for example question **5.** (b), 2013.

> In a combustion chamber, cyanogen gas burns to form a mixture of carbon dioxide and nitrogen.
>
> $$C_2N_2(g) + 2O_2(g) \rightarrow 2CO_2(g) + N_2(g)$$
>
> Carbon dioxide can be removed by passing the gas mixture through sodium hydroxide solution.
>
> Complete the diagram to show how carbon dioxide can be removed from the products and the volume of nitrogen gas measured.   **2**

Where you can achieve extra marks: when asked to draw apparatus, try to draw your diagram as neatly as possible. When drawing diagrams showing gases being collected, be careful not to show delivery tubes passing through the sides of other pieces of apparatus. Take care to make sure that apparatus is correctly positioned to collect any rising bubbles of gas.

Common mistakes to avoid are shown below.

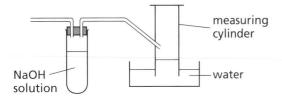

In the diagram above, the tube entering the test tube should extend down into the NaOH solution so that the gas has to bubble through the solution. The delivery tube should not go through the side of the measuring cylinder.

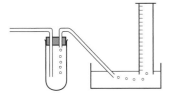

In the diagram above, the apparatus is not shown assembled correctly in order for bubbles of gas rising naturaly through the water to be collected. The bubbles of gas are shown travelling across the tub of water from the delivery tube to the waiting cylinder – an arrangement which would not work.

Problem Solving involving unfamiliar contexts

Each year there is at least one problem-solving question that will present you with information about an unfamiliar area of chemistry (e.g. Question 18, 2013). Although these questions can look intimidating, all of the information you need to answer them is usually contained in the question itself.

Questions requiring a detailed explanation

Each year there will be one or two questions for which you will need to write a detailed explanation. These questions are always signposted for you by the use of the words, "**Explain clearly**" normally printed in bold. (Questions 4(a)(ii) and 6(a), 2013). You can also recognise when a more detailed answer is required by looking at the number of marks allocated to the question. "Explain" questions are usually worth at least two marks. Sometimes the examiners will try to give you additional help by giving you further advice on the things they would expect to see in a good answer.

For example,

> **6.** (a) **Explain clearly** why trichloromethane is more soluble in water than tetrachloromethane. Your answer should include the names of the intermolecular forces involved.   **2**

Many students answering this question correctly stated that trichloromethane was polar whilst tetrachloromethane was non-polar. They also correctly stated that water was a polar solvent and, as like-dissolves-like, trichloromethane would be expected to be soluble in water. Unfortunately, many students overlooked the advice in this question which stated, "your answer should include the names of the intermolecular forces involved" and so lost marks for not mentioning the permanent dipole-permanent dipole attractions in the polar liquids, and the van der Waals' attractions in the tetrachloromethane.

Where you can achieve extra marks: recognise when a detailed explanation is required, and make sure you follow any advice given in the question in order to gain full marks.

Questions involving the formulae of organic compounds

You will be required to answer questions involving molecular formulae (Questions 5(a), 8(a) and 10(a)(ii), 2013).

Common reasons for candidates to lose marks are:

- failure to show all bonds if a full structural formula is requested
- when bonds are not shown connecting clearly to the correct atom in a shortened structural formula

e.g. Question 10(a)(ii)

$$OH - CH_2 - CH_2 - N \begin{array}{c} {}^{CH_2 - CH_3} \\ {}_{CH_2 - CH_3} \end{array}$$

In this example, the hydroxyl group is incorrectly bonded through the hydrogen atom.

Where you can achieve extra marks: be careful when drawing structural formula to make sure that the bonds are joined to the correct atoms.

Good luck!

Remember that the rewards for passing Higher Chemistry are well worth it! Your pass will help you get the future you want for yourself. In the exam, be confident in your own ability, if you're not sure how to answer a question trust your instincts and just give it a go anyway – keep calm and don't panic! GOOD LUCK!

[BLANK PAGE]

FOR OFFICIAL USE

Total
Section B

X012/301

NATIONAL
QUALIFICATIONS
2010

WEDNESDAY, 2 JUNE
9.00 AM – 11.30 AM

CHEMISTRY
HIGHER

Fill in these boxes and read what is printed below.

Full name of centre

Town

Forename(s)

Surname

Date of birth

Day	Month	Year	Scottish candidate number	Number of seat

Reference may be made to the Chemistry Higher and Advanced Higher Data Booklet.

SECTION A   Questions 1 40 (40 marks)

Instructions for completion of **Section A** are given on page two.

For this section of the examination you must use an **HB pencil**.

SECTION B (60 marks)

1   All questions should be attempted.

2   The questions may be answered in any order but all answers are to be written in the spaces provided in this answer book, **and must be written clearly and legibly in ink**.

3   Rough work, if any should be necessary, should be written in this book and then scored through when the fair copy has been written. If further space is required, a supplementary sheet for rough work may be obtained from the Invigilator.

4   Additional space for answers will be found at the end of the book. If further space is required, supplementary sheets may be obtained from the Invigilator and should be inserted inside the **front** cover of this book.

5   The size of the space provided for an answer should not be taken as an indication of how much to write. It is not necessary to use all the space.

6   Before leaving the examination room you must give this book to the Invigilator. If you do not, you may lose all the marks for this paper.

SECTION A

Read carefully

1 Check that the answer sheet provided is for **Chemistry Higher (Section A)**.

2 For this section of the examination you must use an **HB pencil** and, where necessary, an eraser.

3 Check that the answer sheet you have been given has **your name**, **date of birth**, **SCN** (Scottish Candidate Number) and **Centre Name** printed on it.

   Do not change any of these details.

4 If any of this information is wrong, tell the Invigilator immediately.

5 If this information is correct, **print** your name and seat number in the boxes provided.

6 The answer to each question is **either** A, B, C or D. Decide what your answer is, then, using your pencil, put a horizontal line in the space provided (see sample question below).

7 There is **only one correct** answer to each question.

8 Any rough working should be done on the question paper or the rough working sheet, **not** on your answer sheet.

9 At the end of the examination, put the **answer sheet for Section A inside the front cover of your answer book**.

Sample Question

To show that the ink in a ball-pen consists of a mixture of dyes, the method of separation would be

        A chromatography

        B fractional distillation

        C fractional crystallisation

        D filtration.

The correct answer is **A**—chromatography. The answer **A** has been clearly marked in **pencil** with a horizontal line (see below).

Changing an answer

If you decide to change your answer, carefully erase your first answer and using your pencil, fill in the answer you want. The answer below has been changed to **D**.

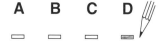

1. Which of the following gases would dissolve in water to form an alkali?

   A   HBr

   B NH_3

   C CO_2

   D CH_4

2. Which of the following pairs of solutions is most likely to produce a precipitate when mixed?

   A   Magnesium nitrate + sodium chloride

   B   Magnesium nitrate + sodium sulphate

   C   Silver nitrate + sodium chloride

   D   Silver nitrate + sodium sulphate

3. 0·5 mol of copper(II) chloride and 0·5 mol of copper(II) sulphate are dissolved together in water and made up to $500\,cm^3$ of solution.

   What is the concentration of Cu^{2+}(aq) ions in the solution in $mol\,l^{-1}$?

   A   0·5

   B   1·0

   C   2·0

   D   4·0

4. For any chemical, its temperature is a measure of

   A   the average kinetic energy of the particles that react

   B   the average kinetic energy of all the particles

   C   the activation energy

   D   the minimum kinetic energy required before reaction occurs.

5. 1 mol of hydrogen gas and 1 mol of iodine vapour were mixed and allowed to react. After t seconds, 0·8 mol of hydrogen remained.

   The number of moles of hydrogen iodide formed at t seconds was

   A   0·2

   B   0·4

   C   0·8

   D   1·6.

6. Excess zinc was added to $100\,cm^3$ of hydrochloric acid, concentration $1\,mol\,l^{-1}$.

   Graph I refers to this reaction.

   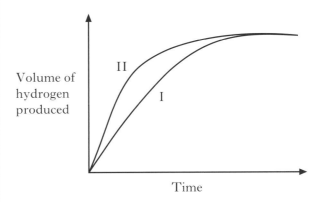

   Graph II could be for

   A   excess zinc reacting with $100\,cm^3$ of hydrochloric acid, concentration $2\,mol\,l^{-1}$

   B   excess zinc reacting with $100\,cm^3$ of sulphuric acid, concentration $1\,mol\,l^{-1}$

   C   excess zinc reacting with $100\,cm^3$ of ethanoic acid, concentration $1\,mol\,l^{-1}$

   D   excess magnesium reacting with $100\,cm^3$ of hydrochloric acid, concentration $1\,mol\,l^{-1}$.

7. Which of the following is **not** a correct statement about the effect of a catalyst?

   The catalyst

   A   provides an alternative route to the products

   B   lowers the energy that molecules need for successful collisions

   C   provides energy so that more molecules have successful collisions

   D   forms bonds with reacting molecules.

8. A potential energy diagram can be used to show the activation energy (E_A) and the enthalpy change (ΔH) for a reaction.

   Which of the following combinations of E_A and ΔH could **never** be obtained for a reaction?

   A $E_A = 50\,kJ\,mol^{-1}$ and $\Delta H = -100\,kJ\,mol^{-1}$

   B $E_A = 50\,kJ\,mol^{-1}$ and $\Delta H = +100\,kJ\,mol^{-1}$

   C $E_A = 100\,kJ\,mol^{-1}$ and $\Delta H = +50\,kJ\,mol^{-1}$

   D $E_A = 100\,kJ\,mol^{-1}$ and $\Delta H = -50\,kJ\,mol^{-1}$

[Turn over

9. As the relative atomic mass in the halogens increases

   A the boiling point increases

   B the density decreases

   C the first ionisation energy increases

   D the atomic size decreases.

10. The table shows the first three ionisation energies of aluminium.

Ionisation energy/kJ mol^{-1}		
1st	2nd	3rd
584	1830	2760

Using this information, what is the enthalpy change, in kJ mol^{-1}, for the following reaction?

$$Al^{3+}(g) + 2e^- \rightarrow Al^+(g)$$

   A +2176

   B −2176

   C +4590

   D −4590

11. When two atoms form a non-polar covalent bond, the two atoms **must** have

   A the same atomic size

   B the same electronegativity

   C the same ionisation energy

   D the same number of outer electrons.

12. In which of the following liquids does hydrogen bonding occur?

   A Ethanoic acid

   B Ethyl ethanoate

   C Hexane

   D Hex-1-ene

13. Which line in the table shows the correct entries for tetrafluoroethene?

	Polar bonds?	Polar molecules?
A	yes	yes
B	yes	no
C	no	no
D	no	yes

14. Element **X** was found to have the following properties.

   (i) It does not conduct electricity when solid.

   (ii) It forms a gaseous oxide.

   (iii) It is a solid at room temperature.

Element **X** could be

   A magnesium

   B silicon

   C nitrogen

   D sulphur.

15. The Avogadro Constant is the same as the number of

   A molecules in 16 g of oxygen

   B ions in 1 litre of sodium chloride solution, concentration 1 mol l^{-1}

   C atoms in 24 g of carbon

   D molecules in 2 g of hydrogen.

16. Which of the following contains one mole of neutrons?

   A 1 g of $^{1}_{1}H$

   B 1 g of $^{12}_{6}C$

   C 2 g of $^{24}_{12}Mg$

   D 2 g of $^{22}_{10}Ne$

17. $20 \, cm^3$ of ammonia gas reacted with an excess of heated copper(II) oxide.

$$3CuO + 2NH_3 \rightarrow 3Cu + 3H_2O + N_2$$

Assuming all measurements were made at $200 \, ^\circ C$, what would be the volume of gaseous products?

A $10 \, cm^3$

B $20 \, cm^3$

C $30 \, cm^3$

D $40 \, cm^3$

18. Which of the following fuels can be produced by the fermentation of biological material under anaerobic conditions?

A Methane

B Ethane

C Propane

D Butane

19. Rum flavouring is based on the compound with the formula shown.

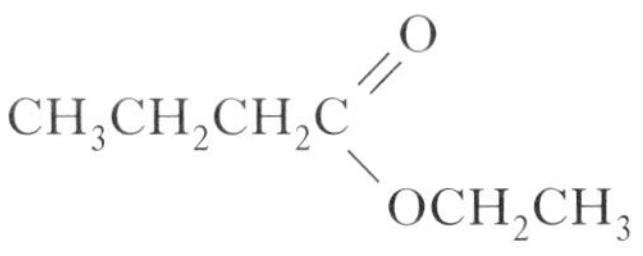

It can be made from

A ethanol and butanoic acid

B propanol and ethanoic acid

C butanol and methanoic acid

D propanol and propanoic acid.

20. Which of the following structural formulae represents a tertiary alcohol?

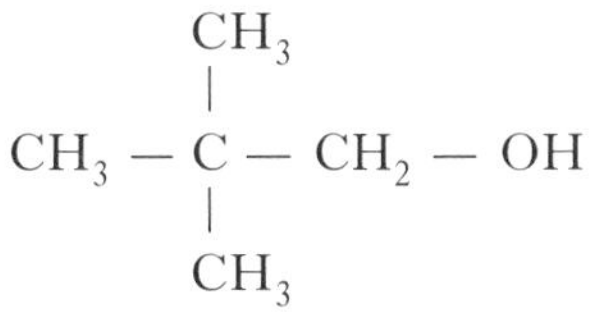

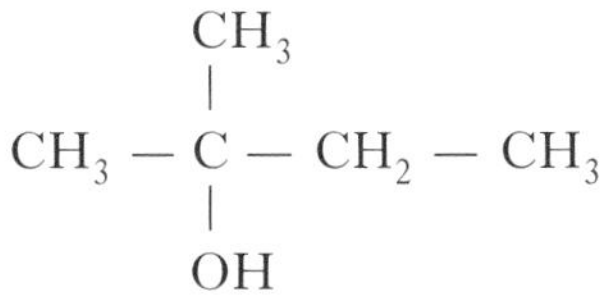

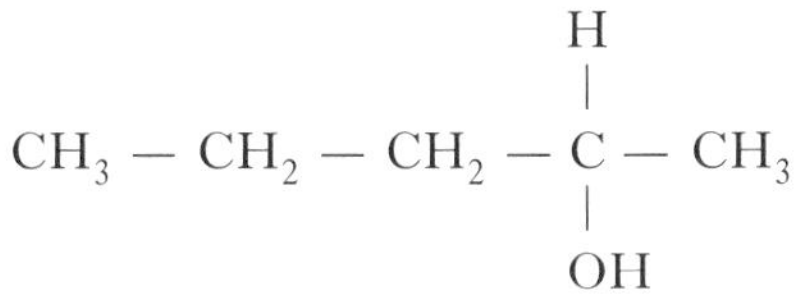

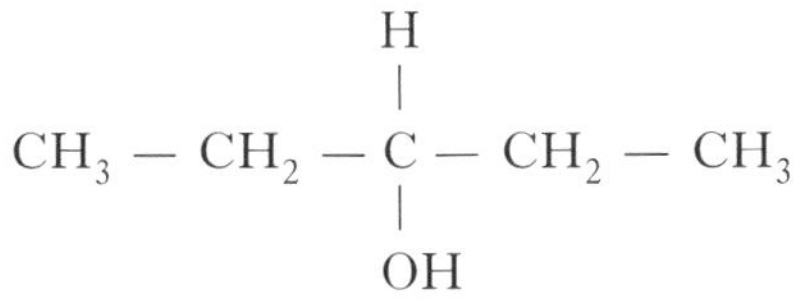

21. What is the product when one mole of chlorine gas reacts with one mole of ethyne?

A 1,1-Dichloroethene

B 1,1-Dichloroethane

C 1,2-Dichloroethene

D 1,2-Dichloroethane

[Turn over

22.

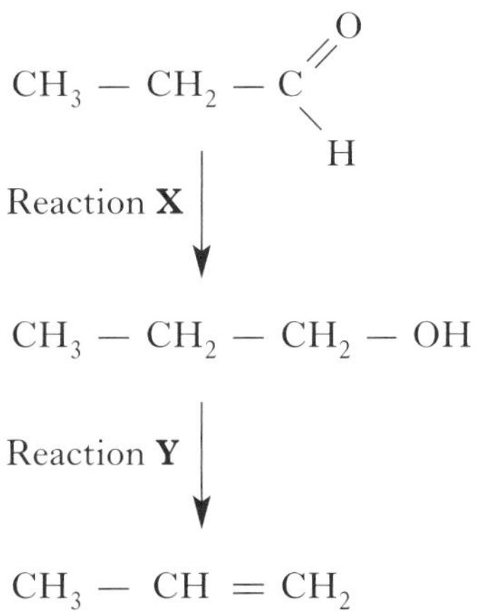

$$CH_3 - CH_2 - C{\Large\diagup}^{\displaystyle O}_{\displaystyle H}$$

Reaction **X** $\downarrow$

$$CH_3 - CH_2 - CH_2 - OH$$

Reaction **Y** $\downarrow$

$$CH_3 - CH = CH_2$$

Which line in the table correctly describes reactions **X** and **Y**?

	Reaction X	Reaction Y
A	oxidation	dehydration
B	oxidation	condensation
C	reduction	dehydration
D	reduction	condensation

23. Ozone has an important role in the upper atmosphere because it

A   absorbs ultraviolet radiation

B   absorbs certain CFCs

C   reflects ultraviolet radiation

D   reflects certain CFCs.

24. Synthesis gas consists mainly of

A CH_4 alone

B CH_4 and CO

C CO and H_2

D CH_4, CO and H_2.

25. Ethene is used in the manufacture of addition polymers.

What type of reaction is used to produce ethene from ethane?

A   Cracking

B   Addition

C   Oxidation

D   Hydrogenation

26. Polyester fibres and cured polyester resins are both very strong.

Which line in the table correctly describes the structure of these polyesters?

	Fibre	Cured resin
A	cross-linked	cross-linked
B	linear	linear
C	cross-linked	linear
D	linear	cross-linked

27. Part of a polymer chain is shown below.

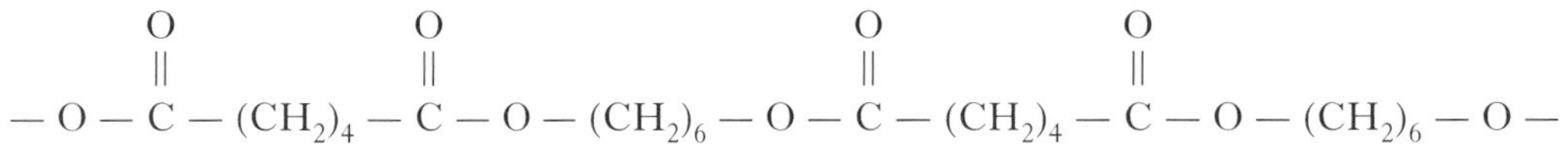

$$- O - \overset{\displaystyle O}{\overset{\displaystyle \|}{C}} - (CH_2)_4 - \overset{\displaystyle O}{\overset{\displaystyle \|}{C}} - O - (CH_2)_6 - O - \overset{\displaystyle O}{\overset{\displaystyle \|}{C}} - (CH_2)_4 - \overset{\displaystyle O}{\overset{\displaystyle \|}{C}} - O - (CH_2)_6 - O -$$

Which of the following compounds, when added to the reactants during polymerisation, would stop the polymer chain from getting too long?

A $HO - \overset{\displaystyle O}{\overset{\displaystyle \|}{C}} - (CH_2)_4 - \overset{\displaystyle O}{\overset{\displaystyle \|}{C}} - OH$

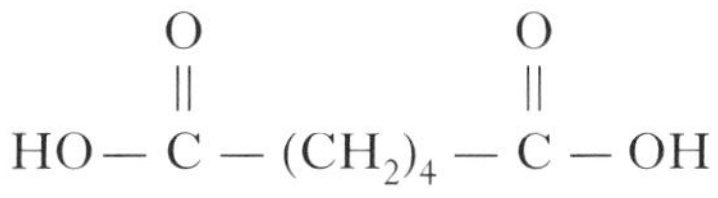

B $HO - (CH_2)_6 - OH$

C $HO - (CH_2)_5 - \overset{\displaystyle O}{\overset{\displaystyle \|}{C}} - OH$

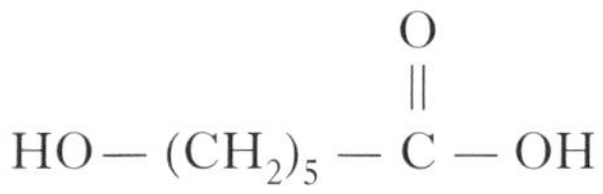

D $CH_3 - (CH_2)_4 - CH_2 - OH$

28. Which of the following fatty acids is unsaturated?

  A $C_{19}H_{39}COOH$

  B $C_{21}H_{43}COOH$

  C $C_{17}H_{31}COOH$

  D $C_{13}H_{27}COOH$

29. Which of the following alcohols is likely to be obtained on hydrolysis of butter?

  A $CH_3 — CH_2 — CH_2 — OH$

  B $CH_3 — \underset{\underset{OH}{|}}{CH} — CH_3$

  C $\begin{array}{l} CH_2 — OH \\ | \\ CH_2 \\ | \\ CH_2 — OH \end{array}$

  D $\begin{array}{l} CH_2 — OH \\ | \\ CH — OH \\ | \\ CH_2 — OH \end{array}$

30. Amino acids are converted into proteins by

  A hydration

  B hydrolysis

  C hydrogenation

  D condensation.

31. Which of the following compounds is a raw material in the chemical industry?

  A Ammonia

  B Calcium carbonate

  C Hexane

  D Nitric acid

32. Given the equations

  $Mg(s) + 2H^+(aq) \rightarrow Mg^{2+}(aq) + H_2(g)$
  $\Delta H = a\ J\ mol^{-1}$

  $Zn(s) + 2H^+(aq) \rightarrow Zn^{2+}(aq) + H_2(g)$
  $\Delta H = b\ J\ mol^{-1}$

  $Mg(s) + Zn^{2+}(aq) \rightarrow Mg^{2+}(aq) + Zn(s)$
  $\Delta H = c\ J\ mol^{-1}$

  then, according to Hess's Law

  A $c = a - b$

  B $c = a + b$

  C $c = b - a$

  D $c = -b - a.$

33. In which of the following reactions would an increase in pressure cause the equilibrium position to move to the left?

  A $CO(g) + H_2O(g) \rightleftharpoons CO_2(g) + H_2(g)$

  B $CH_4(g) + H_2O(g) \rightleftharpoons CO(g) + 3H_2(g)$

  C $Fe_2O_3(s) + 3CO(g) \rightleftharpoons 2Fe(s) + 3CO_2(g)$

  D $N_2(g) + 3H_2(g) \rightleftharpoons 2NH_3(g)$

34. If ammonia is added to a solution containing copper(II) ions an equilibrium is set up.

  $Cu^{2+}(aq) + 2OH^-(aq) + 4NH_3(aq) \rightleftharpoons Cu(NH_3)_4(OH)_2(aq)$
  (deep blue)

  If acid is added to this equilibrium system

  A the intensity of the deep blue colour will increase

  B the equilibrium position will move to the right

  C the concentration of $Cu^{2+}(aq)$ ions will increase

  D the equilibrium position will not be affected.

35. Which of the following is the best description of a $0 \cdot 1\ mol\ l^{-1}$ solution of hydrochloric acid?

  A Dilute solution of a weak acid

  B Dilute solution of a strong acid

  C Concentrated solution of a weak acid

  D Concentrated solution of a strong acid

[Turn over

36. A solution has a negative pH value.

    This solution

    A   neutralises $H^+(aq)$ ions

    B   contains no $OH^-(aq)$ ions

    C   has a high concentration of $H^+(aq)$ ions

    D   contains neither $H^+(aq)$ ions nor $OH^-(aq)$ ions.

37. When a certain aqueous solution is diluted, its conductivity decreases but its pH remains constant.

    It could be

    A   ethanoic acid

    B   sodium chloride

    C   sodium hydroxide

    D   nitric acid.

38. Equal volumes of four $1 \, mol \, l^{-1}$ solutions were compared.

    Which of the following $1 \, mol \, l^{-1}$ solutions contains the most ions?

    A   Nitric acid

    B   Hydrochloric acid

    C   Ethanoic acid

    D   Sulphuric acid

39. In which reaction is hydrogen gas acting as an oxidising agent?

    A $H_2 + CuO \rightarrow H_2O + Cu$

    B $H_2 + C_2H_4 \rightarrow C_2H_6$

    C $H_2 + Cl_2 \rightarrow 2HCl$

    D $H_2 + 2Na \rightarrow 2NaH$

40. Which particle will be formed when an atom of $^{211}_{83}Bi$ emits an α-particle and the decay product then emits a β-particle?

    A $^{207}_{82}Pb$

    B $^{208}_{81}Tl$

    C $^{209}_{80}Hg$

    D $^{210}_{79}Au$

Candidates are reminded that the answer sheet MUST be returned INSIDE the front cover of this answer book.

DO NOT WRITE IN THIS MARGIN

Marks

SECTION B

All answers must be written clearly and legibly in ink.

1. The elements lithium, boron and nitrogen are in the second period of the Periodic Table.

   Complete the table below to show **both** the bonding and structure of these three elements at room temperature.

Name of element	Bonding	Structure
lithium		lattice
boron		
nitrogen	covalent	

2

(2)

[Turn over

Marks

2. (a) Polyhydroxyamide is a recently developed fire-resistant polymer.

The monomers used to produce the polymer are shown.

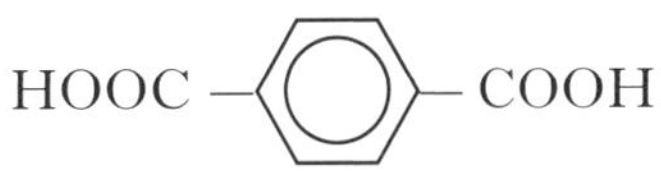

HOOC —⬡— COOH   H$_2$N —⬡— NH$_2$
                                         |
                                         OH

        diacid 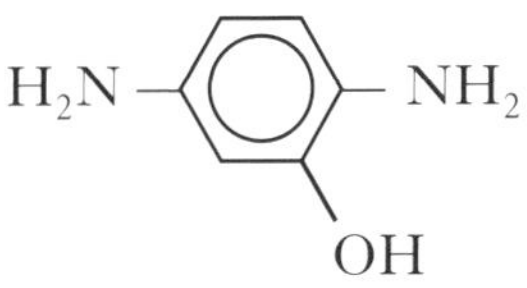   diamine

(i) How many hydrogen atoms are present in a molecule of the diamine molecule?

1

(ii) Draw a section of polyhydroxyamide showing **one** molecule of each monomer joined together.

1

(b) Poly(ethenol), another recently developed polymer, has an unusual property for a plastic.

What is this unusual property?

1

(3)

DO NOT WRITE IN THIS MARGIN

Marks

3. Atmospheric oxygen, $O_2(g)$, dissolves in the Earth's oceans forming dissolved oxygen, $O_2(aq)$, which is essential for aquatic life.

An equilibrium is established.

$$O_2(g) \quad + \quad (aq) \quad \rightleftharpoons \quad O_2(aq) \qquad \Delta H = -12{\cdot}1\ \text{kJ mol}^{-1}$$

(*a*) (i) What is meant by a reaction at "equilibrium"? **1**

(ii) What would happen to the concentration of dissolved oxygen if the temperature of the Earth's oceans increased? **1**

(*b*) A sample of oceanic water was found to contain 0·010 g of dissolved oxygen.

Calculate the number of moles of dissolved oxygen present in the sample. **1**

(3)

Marks

4. In the Hall-Heroult Process, aluminium is produced by the electrolysis of an ore containing aluminium oxide.

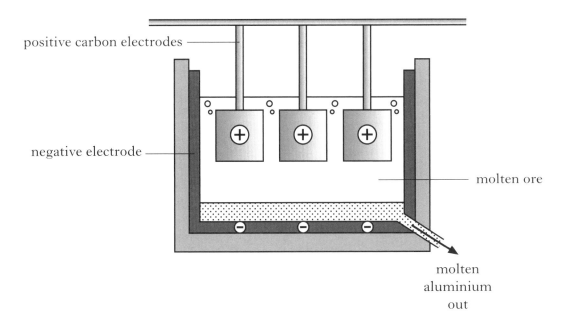

(a) Suggest why the positive carbon electrodes need to be replaced regularly.

1

(b) Calculate the mass of aluminium, in grams, produced in 20 minutes when a current of 50 000 A is used.

Show your working clearly.

3

(4)

DO NOT
WRITE IN
THIS
MARGIN

Marks

5. The reaction of oxalic acid with an acidified solution of potassium permanganate was studied to determine the effect of temperature changes on reaction rate.

$$5(COOH)_2(aq) + 6H^+(aq) + 2MnO_4^-(aq) \rightarrow 2Mn^{2+}(aq) + 10CO_2(g) + 8H_2O(\ell)$$

The reaction was carried out at several temperatures between 40 °C and 60 °C. The end of the reaction was indicated by a colour change from purple to colourless.

(*a*)   (i) State **two** factors that should be kept the same in these experiments.

1

(ii) Why is it difficult to measure an accurate value for the reaction time when the reaction is carried out at room temperature?

1

(*b*)   Sketch a graph to show how the rate varied with increasing temperature.

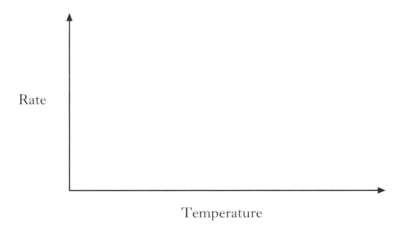

Rate

Temperature

1

(3)

[Turn over

Marks

6. Positron emission tomography, PET, is a technique that provides information about biochemical processes in the body.

Carbon-11, ^{11}C, is a positron-emitting radioisotope that is injected into the bloodstream.

A positron can be represented as $^{0}_{1}e$.

(a) Complete the nuclear equation for the decay of ^{11}C by positron-emission.

$$^{11}C \longrightarrow$$

1

(b) A sample of ^{11}C had an initial count rate of 640 counts min^{-1}. After 1 hour the count rate had fallen to 80 counts min^{-1}.

Calculate the half-life, in minutes, of ^{11}C.

1

(c) ^{11}C is injected into the bloodstream as glucose molecules ($C_6H_{12}O_6$). Some of the carbon atoms in these glucose molecules are ^{11}C atoms.

The intensity of radiation in a sample of ^{11}C is compared with the intensity of radiation in a sample of glucose containing ^{11}C atoms. Both samples have the same mass.

Which sample has the higher intensity of radiation?

Give a reason for your answer.

1

(3)

Marks

7.   Hydrogen cyanide, HCN, is highly toxic.

(*a*)   Information about hydrogen cyanide is given in the table.

Structure	$H-C\equiv N$
Molecular mass	27
Boiling point	$26\,^{\circ}C$

Although hydrogen cyanide has a similar molecular mass to nitrogen, it has a much higher boiling point. This is due to the permanent dipole–permanent dipole attractions in liquid hydrogen cyanide.

What is meant by permanent dipole–permanent dipole attractions?

Explain how they arise in liquid hydrogen cyanide.

2

(*b*)   Hydrogen cyanide is of great importance in organic chemistry. It offers a route to increasing the chain length of a molecule.

If ethanal is reacted with hydrogen cyanide and the product hydrolysed with acid, lactic acid is formed.

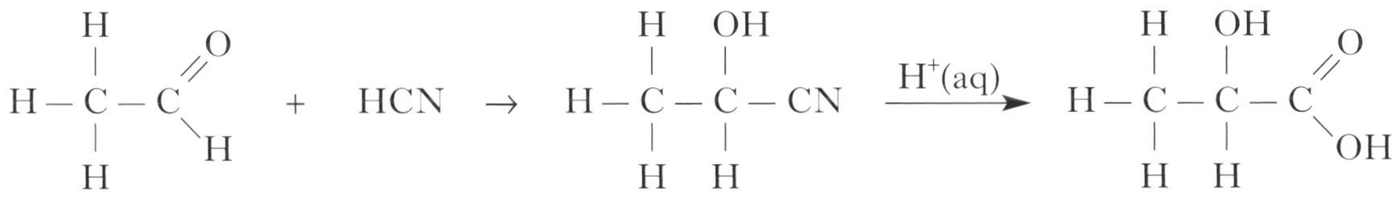

ethanal   lactic acid

Draw a structural formula for the acid produced when propanone is used instead of ethanal in the above reaction sequence.

1

(3)

Marks

8. Glycerol, $C_3H_8O_3$, is widely used as an ingredient in toothpaste and cosmetics.

(*a*) Glycerol is mainly manufactured from fats and oils. Propene can be used as a feedstock in an alternative process as shown.

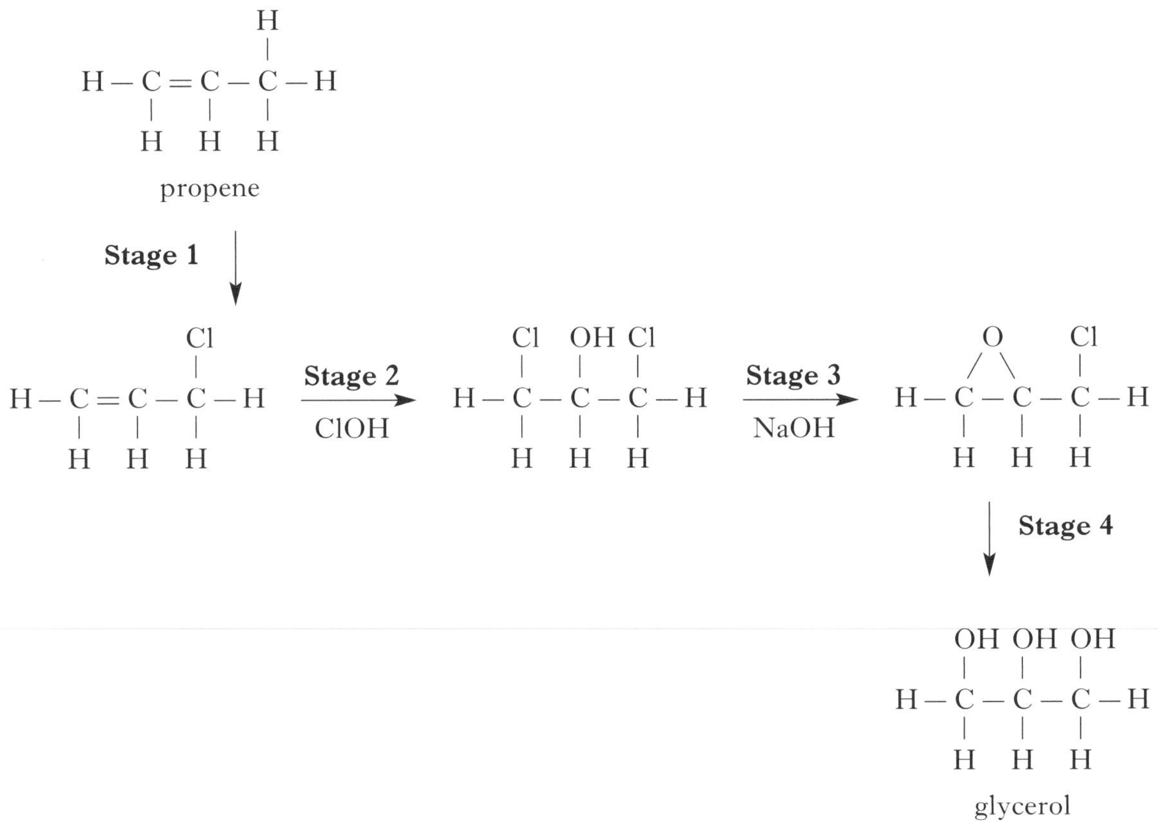

```
            H
            |
    H — C = C — C — H
        |   |   |
        H   H   H
          propene
```

Stage 1 ↓

```
        Cl                          Cl  OH  Cl                         O       Cl
        |                           |   |   |                         / \      |
H — C = C — C — H    Stage 2    H — C — C — C — H    Stage 3    H — C — C — C — H
    |   |   |       ─────────       |   |   |       ─────────       |   |   |
    H   H   H         ClOH          H   H   H         NaOH          H   H   H
```

Stage 4 ↓

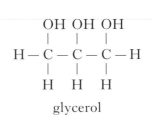

```
        OH  OH  OH
        |   |   |
H — C — C — C — H
    |   |   |
    H   H   H
       glycerol
```

(i) What is meant by a feedstock?

1

(ii) Name the type of reaction taking place in **Stage 2**.

1

(iii) In **Stage 3**, a salt and water are produced as by-products.

Name the salt produced.

1

Marks

8. **(a)** **(continued)**

   (iv) Apart from cost, state **one** advantage of using fats and oils rather than propene in the manufacture of glycerol.

   1

   (b) Hydrogen has been named as a 'fuel for the future'. In a recent article researchers reported success in making hydrogen from glycerol:

   $$C_3H_8O_3(\ell) \rightarrow CO_2(g) + CH_4(g) + H_2(g)$$

   Balance this equation.

   1

   (c) The enthalpy of formation of glycerol is the enthalpy change for the reaction:

   $$3C(s) + 4H_2(g) + 1\frac{1}{2}O_2(g) \rightarrow C_3H_8O_3(\ell)$$
   (graphite)

   Calculate the enthalpy of formation of glycerol, in $kJ\,mol^{-1}$, using information from the data booklet and the following data.

   $$C_3H_8O_3(\ell) + 3\frac{1}{2}O_2(g) \rightarrow 3CO_2(g) + 4H_2O(\ell) \quad \Delta H = -1654\,kJ\,mol^{-1}$$

   Show your working clearly.

   2

   (7)

[Turn over

DO NOT WRITE IN THIS MARGIN

Marks

9. Enzymes are biological catalysts.

(*a*) Name the **four** elements present in all enzymes. **1**

(*b*) The enzyme catalase, found in potatoes, can catalyse the decomposition of hydrogen peroxide.

$$2H_2O_2(aq) \rightarrow 2H_2O(\ell) + O_2(g)$$

A student carried out the Prescribed Practical Activity (PPA) to determine the effect of pH on enzyme activity.

Describe how the activity of the enzyme was measured in this PPA. **1**

(*c*) A student wrote the following **incorrect** statement.

When the temperature is increased, enzyme-catalysed reactions will always speed up because more molecules have kinetic energy greater than the activation energy.

Explain the mistake in the student's reasoning. **1**

(3)

Marks

10. Sulphur trioxide can be prepared in the laboratory by the reaction of sulphur dioxide with oxygen.

$$2SO_2(g) + O_2(g) \rightleftharpoons 2SO_3(g)$$

The sulphur dioxide and oxygen gases are dried by bubbling them through concentrated sulphuric acid. The reaction mixture is passed over heated vanadium(V) oxide.

Sulphur trioxide has a melting point of $17\,^{\circ}C$. It is collected as a white crystalline solid.

(a) Complete the diagram to show how the reactant gases are dried and the product is collected.

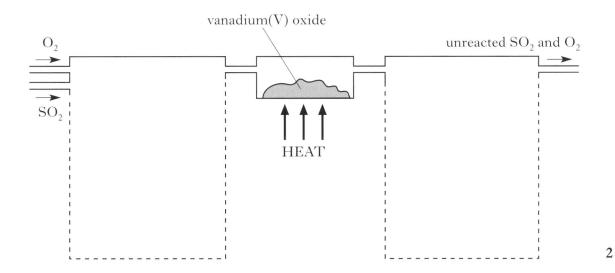

(b) Under certain conditions, 43·2 tonnes of sulphur trioxide are produced in the reaction of 51·2 tonnes of sulphur dioxide with excess oxygen.

Calculate the percentage yield of sulphur trioxide.

Show your working clearly.

2

2

(4)

[Turn over

Marks

11. (*a*) The first ionisation energy of an element is defined as the energy required to remove one mole of electrons from one mole of atoms in the gaseous state.

The graph shows the first ionisation energies of the Group 1 elements.

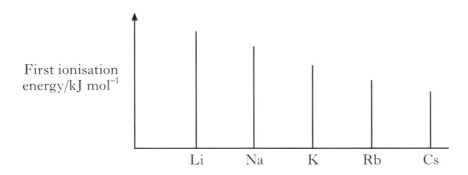

(i) Clearly explain why the first ionisation energy decreases down this group.

2

(ii) The energy needed to remove one electron from one helium atom is $3 \cdot 94 \times 10^{-21}$ kJ.

Calculate the first ionisation energy of helium, in kJ mol^{-1}.

1

(*b*) The ability of an atom to form a negative ion is measured by its Electron Affinity.

The Electron Affinity is defined as the energy change when one mole of gaseous atoms of an element combines with one mole of electrons to form gaseous negative ions.

Write the equation, showing state symbols, that represents the Electron Affinity of chlorine.

1

(4)

DO NOT WRITE IN THIS MARGIN

Marks

12. (*a*) A student bubbled 240 cm^3 of carbon dioxide into 400 cm^3 of 0·10 mol l^{-1} lithium hydroxide solution.

The equation for the reaction is:

$$2LiOH(aq) + CO_2(g) \rightarrow Li_2CO_3(aq) + H_2O(\ell)$$

Calculate the number of moles of lithium hydroxide that would **not** have reacted.

(Take the molar volume of carbon dioxide to be 24 litres mol^{-1}.)

Show your working clearly.

2

(*b*) What is the pH of the 0·10 mol l^{-1} lithium hydroxide solution used in the experiment?

1

(*c*) Explain why lithium carbonate solution has a pH greater than 7.

In your answer you should mention the **two** equilibria involved.

2

(5)

Marks

13. (*a*) A sample of petrol was analysed to identify the hydrocarbons present. The results are shown in the table.

Number of carbon atoms per molecule	Hydrocarbons present in the sample
4	2-methylpropane
5	2-methylbutane
6	2,3-dimethylbutane
7	2,2-dimethylpentane 2,2,3-trimethylbutane

(i) Draw a structural formula for 2,2,3-trimethylbutane.

1

(ii) The structures of the hydrocarbons in the sample are similar in a number of ways.

What similarity in structure makes these hydrocarbons suitable for use in unleaded petrol?

1

(*b*) In some countries, organic compounds called 'oxygenates' are added to unleaded petrol.

One such compound is MTBE.

$$MTBE \qquad H_3C - \underset{\underset{CH_3}{|}}{\overset{\overset{CH_3}{|}}{C}} - O - CH_3$$

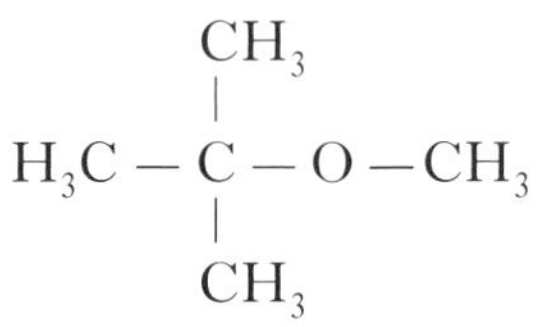

(i) Suggest why oxygenates such as MTBE are added to unleaded petrol.

1

Marks

13. *(b)* **(continued)**

(ii) MTBE is an example of an ether. All ethers contain the functional group:

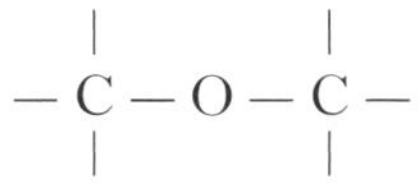

$$-\overset{|}{\underset{|}{C}}-O-\overset{|}{\underset{|}{C}}-$$

Draw a structural formula for an isomer of MTBE that is also an ether.

1

(c) Some of the hydrocarbons that are suitable for unleaded petrol are produced by a process known as reforming.

One reforming reaction is:

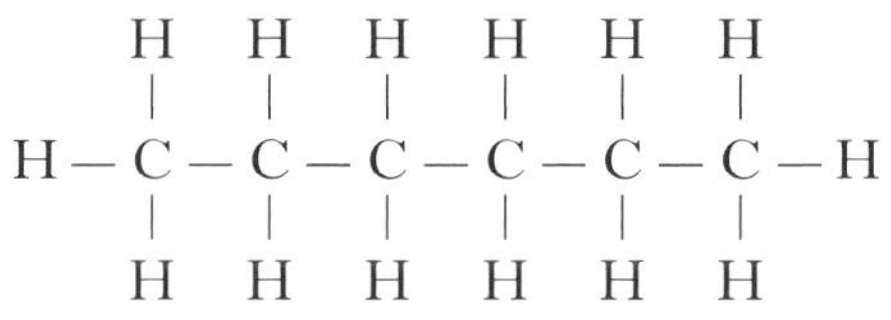

$$H-\overset{\overset{\displaystyle H}{|}}{\underset{\underset{\displaystyle H}{|}}{C}}-\overset{\overset{\displaystyle H}{|}}{\underset{\underset{\displaystyle H}{|}}{C}}-\overset{\overset{\displaystyle H}{|}}{\underset{\underset{\displaystyle H}{|}}{C}}-\overset{\overset{\displaystyle H}{|}}{\underset{\underset{\displaystyle H}{|}}{C}}-\overset{\overset{\displaystyle H}{|}}{\underset{\underset{\displaystyle H}{|}}{C}}-\overset{\overset{\displaystyle H}{|}}{\underset{\underset{\displaystyle H}{|}}{C}}-H \qquad \rightarrow \text{ hydrocarbon } \mathbf{A} \ + \ H_2$$

hexane

Hydrocarbon **A** is non-aromatic and does **not** decolourise bromine solution.

Give a possible name for hydrocarbon **A**.

1

(5)

[Turn over

Marks

14. (*a*) Hess's Law can be verified using the reactions summarised below.

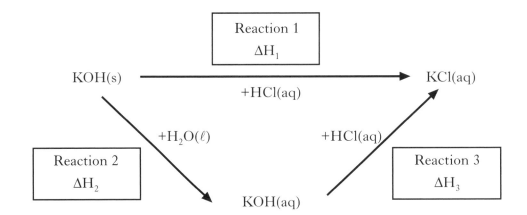

(i) Complete the list of measurements that would have to be carried out in order to determine the enthalpy change for Reaction 2.

Reaction 2

1. Using a measuring cylinder, measure out $25\,cm^3$ of water into a polystyrene cup.

2.

3. Weigh out accurately about $1\cdot2\,g$ of potassium hydroxide and add it to the water, with stirring, until all the solid dissolves.

4.

1

(ii) Why was the reaction carried out in a polystyrene cup?

1

Marks

14. (*a*) (**continued**)

    (iii) A student found that $1{\cdot}08\,kJ$ of energy was **released** when $1{\cdot}2\,g$ of potassium hydroxide was dissolved completely in water.

        Calculate the enthalpy of solution of potassium hydroxide.

1

(*b*) A student wrote the following **incorrect** statement.

The enthalpy of neutralisation for hydrochloric acid reacting with potassium hydroxide is less than that for sulphuric acid reacting with potassium hydroxide because fewer moles of water are formed as shown in these equations.

$$HCl \ + \ KOH \ \rightarrow \ KCl \ + \ H_2O$$

$$H_2SO_4 \ + \ 2KOH \ \rightarrow \ K_2SO_4 \ + \ 2H_2O$$

Explain the mistake in the student's statement.

1
(4)

[Turn over

Marks

15. Infra-red spectroscopy is a technique that can be used to identify the bonds that are present in a molecule.

Different bonds absorb infra-red radiation of different wavenumbers. This is due to differences in the bond 'stretch'. These absorptions are recorded in a spectrum.

A spectrum for propan-1-ol is shown.

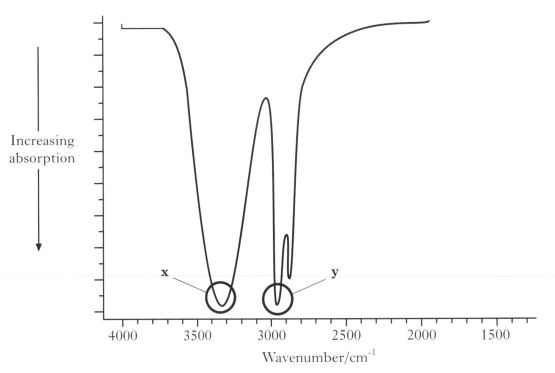

The correlation table on page 13 of the data booklet shows the wavenumber ranges for the absorptions due to different bonds.

(*a*) Use the correlation table to identify the bonds responsible for the two absorptions, **x** and **y**, that are circled in the propan-1-ol spectrum.

    x: **y**:

1

(*b*) Propan-1-ol reacts with ethanoic acid.

    (i) What name is given to this type of reaction?

1

Marks

15. (*b*) **(continued)**

(ii) Draw a spectrum that could be obtained for the organic product of this reaction.

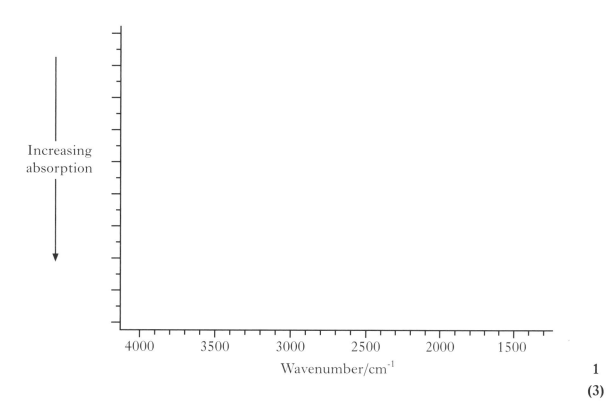

1

(3)

[Turn over

Marks

16. A major problem for the developed world is the pollution of rivers and streams by nitrite and nitrate ions.

    The concentration of nitrite ions, $NO_2^-(aq)$, in water can be determined by titrating samples against acidified permanganate solution.

    (a) Suggest **two** points of good practice that should be followed to ensure that an accurate end-point is achieved in a titration.

    1

    (b) An average of $21\cdot6\,cm^3$ of $0\cdot0150\,mol\,l^{-1}$ acidified permanganate solution was required to react completely with the nitrite ions in a $25\cdot0\,cm^3$ sample of river water.

    The equation for the reaction taking place is:

    $$2MnO_4^-(aq) + 5NO_2^-(aq) + 6H^+(aq) \rightarrow 2Mn^{2+}(aq) + 5NO_3^-(aq) + 3H_2O(\ell)$$

    (i) Calculate the nitrite ion concentration, in $mol\,l^{-1}$, in the river water.

    Show your working clearly.

    2

    (ii) During the reaction the nitrite ion is oxidised to the nitrate ion.

    Complete the ion-electron equation for the oxidation of the nitrite ions.

    $$NO_2^-(aq) \quad \rightarrow \quad NO_3^-(aq)$$

    1

    [END OF QUESTION PAPER]

    (4)

HIGHER

2011

[BLANK PAGE]

FOR OFFICIAL USE

Total
Section B

X012/301

NATIONAL
QUALIFICATIONS
2011

THURSDAY, 26 MAY
9.00 AM – 11.30 AM

CHEMISTRY
HIGHER

Fill in these boxes and read what is printed below.

Full name of centre

Town

Forename(s)

Surname

Date of birth

Day   Month   Year   Scottish candidate number   Number of seat

Reference may be made to the Chemistry Higher and Advanced Higher Data Booklet.

SECTION A—Questions 1–40 (40 marks)

Instructions for completion of **Section A** are given on page two.

For this section of the examination you must use an **HB pencil**.

SECTION B (60 marks)

1   All questions should be attempted.

2   The questions may be answered in any order but all answers are to be written in the spaces provided in this answer book, **and must be written clearly and legibly in ink**.

3   Rough work, if any should be necessary, should be written in this book and then scored through when the fair copy has been written. If further space is required, a supplementary sheet for rough work may be obtained from the Invigilator.

4   Additional space for answers will be found at the end of the book. If further space is required, supplementary sheets may be obtained from the Invigilator and should be inserted inside the **front** cover of this book.

5   The size of the space provided for an answer should not be taken as an indication of how much to write. It is not necessary to use all the space.

6   Before leaving the examination room you must give this book to the Invigilator. If you do not, you may lose all the marks for this paper.

SA X012/301 6/14860

SECTION A

Read carefully

1 Check that the answer sheet provided is for **Chemistry Higher (Section A)**.

2 For this section of the examination you must use an **HB pencil** and, where necessary, an eraser.

3 Check that the answer sheet you have been given has **your name**, **date of birth**, **SCN** (Scottish Candidate Number) and **Centre Name** printed on it.

   Do not change any of these details.

4 If any of this information is wrong, tell the Invigilator immediately.

5 If this information is correct, **print** your name and seat number in the boxes provided.

6 The answer to each question is **either** A, B, C or D. Decide what your answer is, then, using your pencil, put a horizontal line in the space provided (see sample question below).

7 There is **only one correct** answer to each question.

8 Any rough working should be done on the question paper or the rough working sheet, **not** on your answer sheet.

9 At the end of the examination, put the **answer sheet for Section A inside the front cover of your answer book**.

Sample Question

To show that the ink in a ball-pen consists of a mixture of dyes, the method of separation would be

      A chromatography

      B fractional distillation

      C fractional crystallisation

      D filtration.

The correct answer is **A**—chromatography. The answer **A** has been clearly marked in **pencil** with a horizontal line (see below).

Changing an answer

If you decide to change your answer, carefully erase your first answer and using your pencil, fill in the answer you want. The answer below has been changed to **D**.

1. Which of the following gases could be described as monatomic?

   A Bromine

   B Methane

   C Hydrogen

   D Helium

2. Different isotopes of the same element have identical

   A electron arrangements

   B nuclei

   C numbers of neutrons

   D mass numbers.

3. Which of the following pairs of solutions will **not** react to produce a precipitate?

   A Copper(II) chloride and lithium sulphate

   B Potassium carbonate and zinc sulphate

   C Silver nitrate and sodium chloride

   D Ammonium phosphate and magnesium chloride

4. Which of the following sugars does **not** react with Benedict's solution?

   A Glucose

   B Fructose

   C Maltose

   D Sucrose

5. Which of the following gases would contain the **greatest** number of molecules in a 100 g sample, at room temperature?

   A Fluorine

   B Hydrogen

   C Nitrogen

   D Oxygen

6.
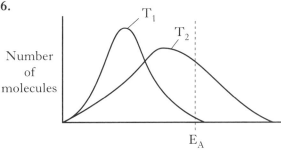

Kinetic energy of molecules

Which line in the table is correct for a reaction as the temperature **decreases** from T_2 to T_1?

	Activation energy (E_A)	Number of successful collisions
A	remains the same	increases
B	decreases	decreases
C	decreases	increases
D	remains the same	decreases

7. A pupil added 0·1 mol of zinc to a solution containing 0·05 mol of silver(I) nitrate.

   $$Zn(s) + 2AgNO_3(aq) \rightarrow Zn(NO_3)_2(aq) + 2Ag(s)$$

   Which of the following statements about the experiment is correct?

   A 0·05 mol of zinc reacts.

   B 0·05 mol of silver is displaced.

   C Silver nitrate is in excess.

   D All of the zinc reacts.

8. A reaction takes place in two stages.

Stage 1

$S_2O_8{}^{2-}(aq) + 2I^-(aq) + 2Fe^{2+}(aq) \rightarrow 2SO_4{}^{2-}(aq) + 2I^-(aq) + 2Fe^{3+}(aq)$

Stage 2

$2SO_4{}^{2-}(aq) + 2I^-(aq) + 2Fe^{3+}(aq) \rightarrow 2SO_4{}^{2-}(aq) + I_2(aq) + 2Fe^{2+}(aq)$

The ion that is the catalyst in the reaction is

   A $S_2O_8{}^{2-}(aq)$

   B $I^-(aq)$

   C $Fe^{2+}(aq)$

   D $SO_4{}^{2-}(aq)$.

9. The following potential diagram is for a reaction carried out with and without a catalyst.

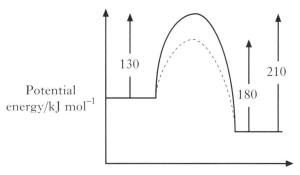

Reaction path

The activation energy for the catalysed reaction is

A $30 \, kJ \, mol^{-1}$

B $80 \, kJ \, mol^{-1}$

C $100 \, kJ \, mol^{-1}$

D $130 \, kJ \, mol^{-1}$.

10. Which of the following equations represents an enthalpy of combustion?

A $C_2H_6(g) + 3\frac{1}{2}O_2(g)$
   $\downarrow$
   $2CO_2(g) + 3H_2O(\ell)$

B $C_2H_5OH(\ell) + O_2(g)$
   $\downarrow$
   $CH_3COOH(\ell) + H_2O(\ell)$

C $CH_3CHO(\ell) + \frac{1}{2}O_2(g)$
   $\downarrow$
   $CH_3COOH(\ell)$

D $CH_4(g) + 1\frac{1}{2}O_2(g)$
   $\downarrow$
   $CO(g) + 2H_2O(\ell)$

11. A potassium atom is larger than a sodium atom because potassium has

A   a larger nuclear charge

B   a larger nucleus

C   more occupied energy levels

D   a smaller ionisation energy.

12. Hydrogen will form a non-polar covalent bond with an element which has an electronegativity value of

A   0·9

B   1·5

C   2·2

D   2·5.

13. Which property of a chloride would prove that it contained ionic bonding?

A   It conducts electricity when molten.

B   It is soluble in a polar solvent.

C   It is a solid at room temperature.

D   It has a high boiling point.

14. A mixture of potassium chloride and potassium carbonate is known to contain 0·1 mol of chloride ions and 0·1 mol of carbonate ions.

How many moles of potassium ions are present?

A   0·15

B   0·20

C   0·25

D   0·30

15. Which of the following has the largest volume under the same conditions of temperature and pressure?

A   1 g hydrogen

B   14 g nitrogen

C   20·2 g neon

D   35·5 g chlorine

16. $20 \, cm^3$ of butane is burned in $150 \, cm^3$ of oxygen.

$C_4H_{10}(g) + 6\frac{1}{2}O_2(g) \rightarrow 4CO_2(g) + 5H_2O(g)$

What is the total volume of gas present after complete combustion of the butane?

A $80 \, cm^3$

B $100 \, cm^3$

C $180 \, cm^3$

D $200 \, cm^3$

17. Which of the following types of hydrocarbons when added to petrol would **not** reduce "knocking"?

    A   Cycloalkanes

    B   Aromatic hydrocarbons

    C   Branched-chain alkanes

    D   Straight-chain alkanes

18. Which of the following fuels when burned would make no contribution to global warming?

    A   Hydrogen

    B   Natural gas

    C   Petrol

    D   Coal

19. Which of the following hydrocarbons always gives the same product when one of its hydrogen atoms is replaced by a chlorine atom?

    A   Hexane

    B   Hex-1-ene

    C   Cyclohexane

    D   Cyclohexene

20.

    The name of this compound is

    A   methanol

    B   methanal

    C   methanoic acid

    D   methanone.

21. Which of the following is an isomer of ethyl propanoate?

    A   Pentan-2-one

    B   Pentanoic acid

    C   Methyl propanoate

    D   Pentane-1,2-diol

22. Propan-2-ol can be prepared from propane as follows.

$$CH_3 - CH_2 - CH_3 \xrightarrow{\text{Step } \mathbf{1}} CH_3 - CH = CH_2 \xrightarrow{\text{Step } \mathbf{2}} CH_3 - \underset{\underset{OH}{|}}{CH} - CH_3$$

Which line in the table correctly describes the types of reaction taking place at Steps **1** and **2**?

	Step 1	Step 2
A	cracking	hydration
B	cracking	hydrolysis
C	dehydration	hydration
D	dehydration	hydrolysis

[**Turn over**

23.

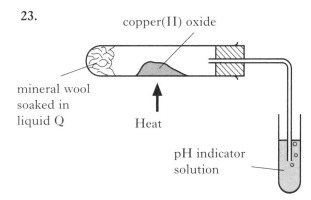

copper(II) oxide

mineral wool
soaked in
liquid Q

Heat

pH indicator
solution

After heating for several minutes, as shown in the diagram, the pH indicator solution turned red.

Liquid **Q** could be

A   propanone

B   paraffin

C   propan-1-ol

D   propan-2-ol.

24. Which of the following compounds is hydrolysed when warmed with sodium hydroxide solution?

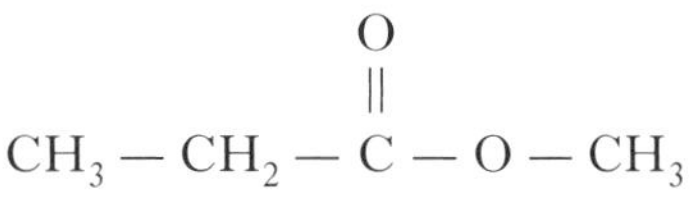

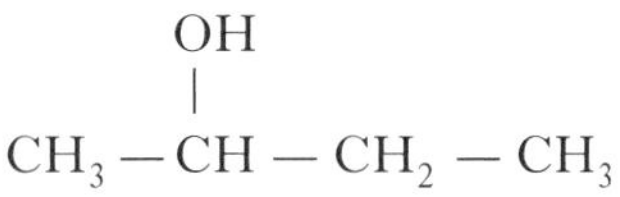

A $CH_3 - \overset{\overset{\text{O}}{\|}}{C} - CH_2 - CH_3$

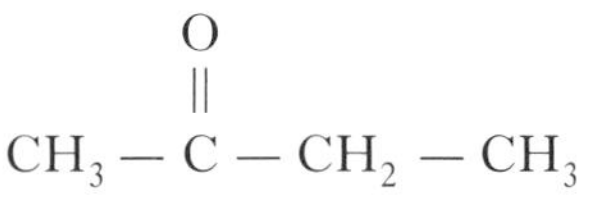

B $CH_3 - CH_2 - \overset{\overset{\text{O}}{\|}}{C} - O - CH_3$

C $CH_3 - \overset{\overset{\text{OH}}{|}}{CH} - CH_2 - CH_3$

D $CH_3 - CH_2 - \underset{\underset{\text{CH}_3}{|}}{CH} - \overset{\overset{\text{O}}{\|}}{C} - H$

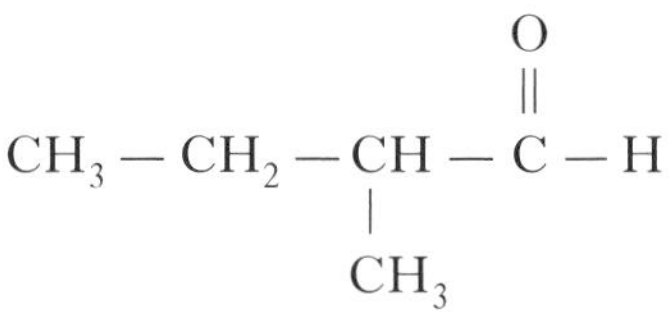

25. Which of the following is most likely to be used as a flavouring?

A CH_3CH_2CHO

B $CH_3CH_2CH_2COOH$

C $CH_3CH(OH)CH_2CH_3$

D $CH_3CH_2CH_2COOCH_2CH_3$

26. Thermosetting plastics can be made by the following sequence of reactions.

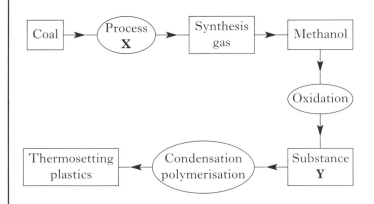

Which line in the table shows the correct names for Process **X** and Substance **Y**?

	Process X	Substance Y
A	combustion	methanal
B	combustion	methanoic acid
C	steam reforming	methanoic acid
D	steam reforming	methanal

27. Which of the following is an amine?

A

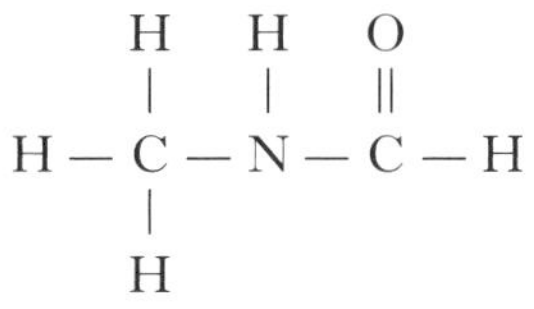

$H - \overset{\overset{\text{H}}{|}}{\underset{\underset{\text{H}}{|}}{C}} - \overset{\overset{\text{H}}{|}}{N} - \overset{\overset{\text{O}}{\|}}{C} - H$

B

$H - \overset{\overset{\text{H}}{|}}{\underset{\underset{\text{H}}{|}}{C}} - \overset{\overset{\text{H}}{|}}{N} - H$

C

$H - \overset{\overset{\text{H}}{|}}{\underset{\underset{\text{H}}{|}}{C}} - N = C = O$

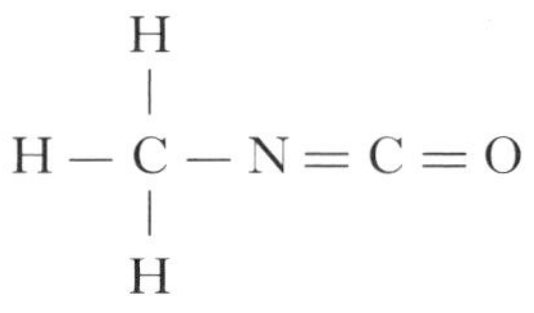

D

$H - \overset{\overset{\text{H}}{|}}{C} = \overset{\overset{\text{H}}{|}}{C} - C \equiv N$

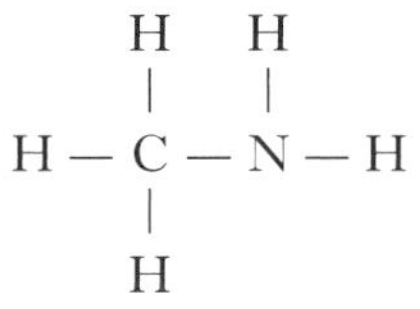

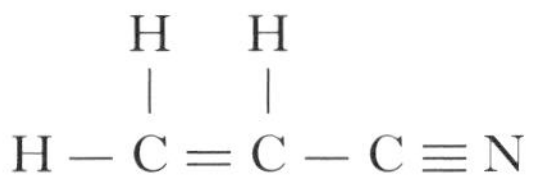

28. Noradrenaline and phenylephrine cause increases in the blood pressure because the part of each of these molecules that they have in common has the correct shape to allow them to bind to a certain human protein.

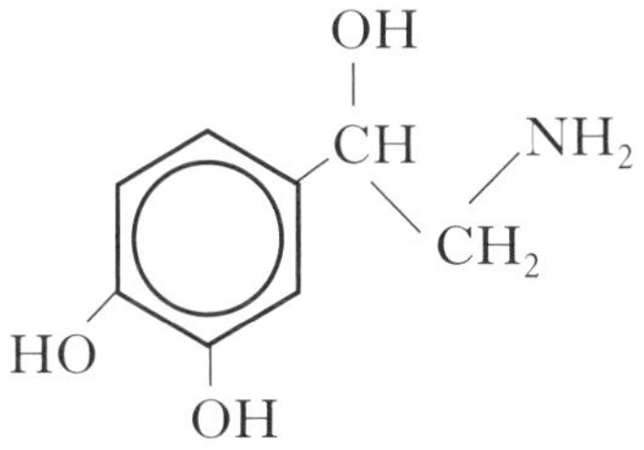

noradrenaline

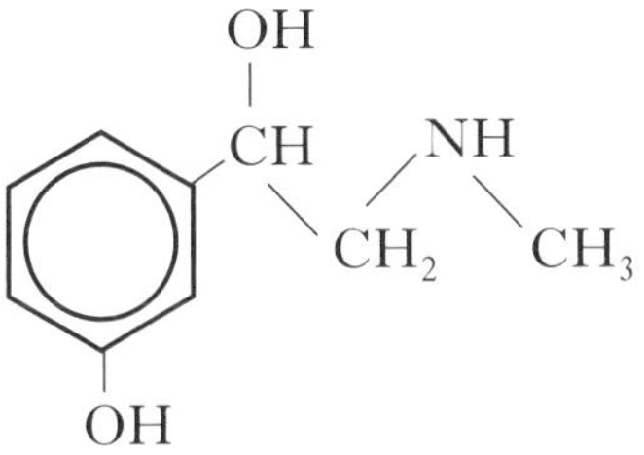

phenylephrine

The part of these molecules which is the correct shape to bind to the protein is

A

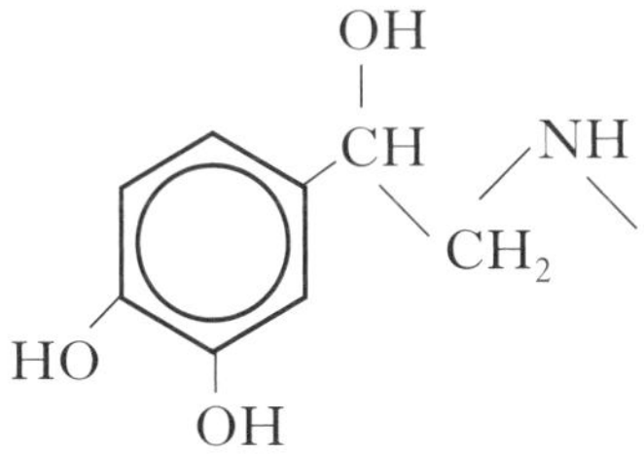

HO

OH

B

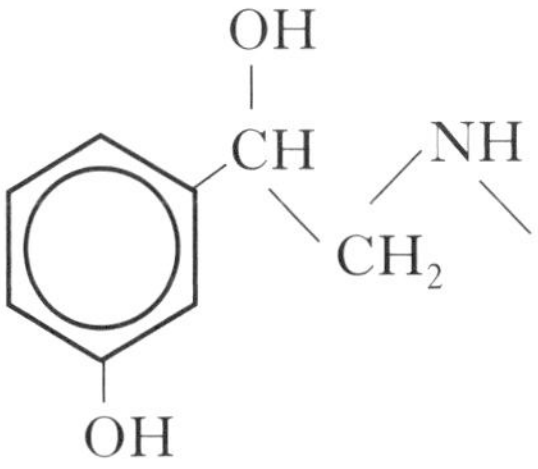

OH

C

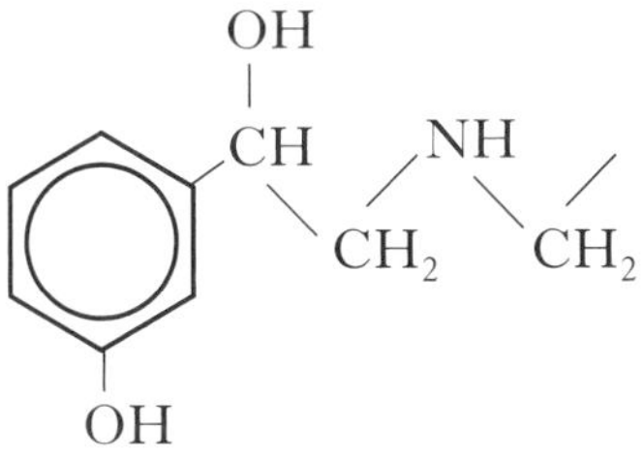

OH

D

HO

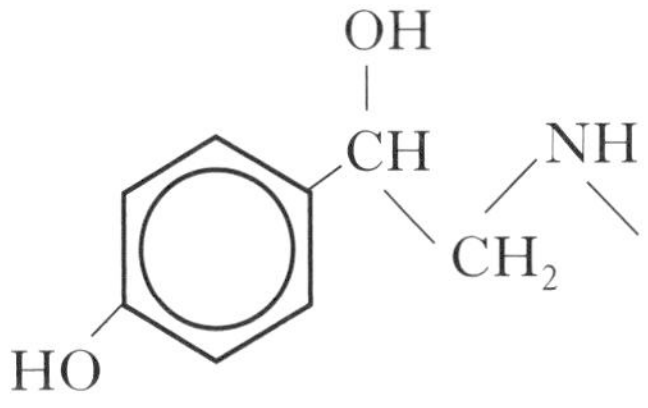

29. Which of the following polymers is photoconductive?

A Kevlar

B Biopol

C Poly(ethenol)

D Poly(vinyl carbazole)

30. Which of the following compounds can be classified as proteins?

A Fats

B Oils

C Enzymes

D Amino acids

31. The flow chart summarises some industrial processes involving ethene.

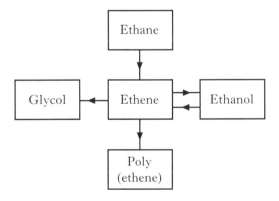

The feedstocks for ethene in these processes are

A ethane and glycol

B ethane and ethanol

C glycol and poly(ethene)

D glycol, poly(ethene) and ethanol.

32. The enthalpy change for $K(s) \rightarrow K(g)$ is $88\,kJ\,mol^{-1}$.

Using the above information and information from the data booklet (page 10), the enthalpy change for $K(s) \rightarrow K^{2+}(g) + 2e^-$ is

A $513\,kJ\,mol^{-1}$

B $3060\,kJ\,mol^{-1}$

C $3485\,kJ\,mol^{-1}$

D $3573\,kJ\,mol^{-1}$.

33. Two flasks, A and B, were placed in a water bath at $40\,°C$.

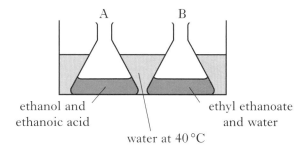

ethanol and
ethanoic acid

ethyl ethanoate
and water

water at $40\,°C$

After several days the contents of both flasks were analysed.

Which result would be expected?

A   Flask A contains ethyl ethanoate, water, ethanol and ethanoic acid; flask B is unchanged.

B   Flask A contains only ethyl ethanoate and water; flask B is unchanged.

C   Flask A contains only ethyl ethanoate and water; flask B contains ethyl ethanoate, water, ethanol and ethanoic acid.

D   Flask A and flask B contain ethyl ethanoate, water, ethanol and ethanoic acid.

34. $NH_3(g) + H_2O(\ell) \rightleftharpoons NH_4^+(aq) + OH^-(aq)$

$\Delta H = -36\,kJ\,mol^{-1}$

The solubility of ammonia in water will be increased by

A   increasing pressure and cooling

B   decreasing pressure and cooling

C   decreasing pressure and warming

D   increasing pressure and warming.

35. The pH of a $0{\cdot}1\,mol\,l^{-1}$ solution of an acid was measured and found to be pH 4.

The pH of a $0{\cdot}001\,mol\,l^{-1}$ solution of an alkali was measured and found to be pH 11.

Which line in the table is correct?

	Acid	Alkali
A	weak	weak
B	weak	strong
C	strong	weak
D	strong	strong

36. Which of the following solutions contains equal concentrations of $H^+(aq)$ and $OH^-(aq)$ ions?

A $NH_4Cl(aq)$

B $Na_2CO_3(aq)$

C $KNO_3(aq)$

D $CH_3COOK(aq)$

37. During a redox process in acid solution, iodate ions are converted into iodine.

$2IO_3^-(aq) + 12H^+(aq) + \mathbf{x}e^- \rightarrow I_2(aq) + 6H_2O(\ell)$

To balance the equation, what is the value of $\mathbf{x}$?

A   2

B   6

C   10

D   12

38. The following reactions take place when nitric acid is added to zinc.

$NO_3^-(aq) + 4H^+(aq) + 3e^- \rightarrow NO(g) + 2H_2O(\ell)$

$Zn(s) \rightarrow Zn^{2+}(aq) + 2e^-$

How many moles of $NO_3^-(aq)$ are reduced by one mole of zinc?

A $\dfrac{2}{3}$

B   1

C $\dfrac{3}{2}$

D   2

39. If 96 500 C of electricity are passed through separate solutions of copper(II) chloride and nickel(II) chloride, then

A equal masses of copper and nickel will be deposited

B the same number of atoms of each metal will be deposited

C the metals will be plated on the positive electrode

D different numbers of moles of each metal will be deposited.

40. Some smoke detectors make use of radiation which is very easily stopped by tiny smoke particles moving between the radioactive source and the detector.

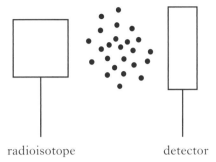

radioisotope   detector

The most suitable type of radioisotope for a smoke detector would be

A an alpha-emitter with a long half-life

B a gamma-emitter with a short half-life

C an alpha-emitter with a short half-life

D a gamma-emitter with a long half-life.

Candidates are reminded that the answer sheet MUST be returned INSIDE the front cover of this answer book.

[Turn over

SECTION B

Marks

All answers must be written clearly and legibly in ink.

1. Chloromethane, CH_3Cl, can be produced by reacting methanol solution with dilute hydrochloric acid using a solution of zinc chloride as a catalyst.

$$CH_3OH(aq) \ + \ HCl(aq) \ \xrightarrow{\ ZnCl_2(aq)\ } \ CH_3Cl(aq) \ + \ H_2O(\ell)$$

(a) What type of catalysis is taking place?

1

(b) The graph shows how the concentration of the hydrochloric acid changed over a period of time when the reaction was carried out at 20 °C.

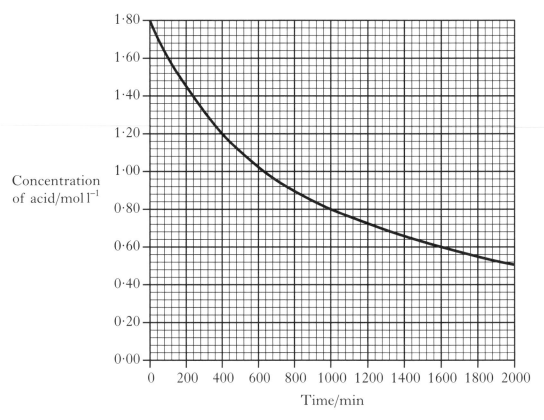

Concentration of acid/mol l^{-1}

Time/min

(i) Calculate the average rate, in mol l^{-1} min^{-1}, in the first 400 minutes.

1

(ii) On the graph above, sketch a curve to show how the concentration of hydrochloric acid would change over time if the reaction is repeated at 30 °C.

(Additional graph paper, if required, can be found on *Page thirty-five*).

1

(3)

Marks

2. The elements from sodium to argon make up the third period of the Periodic Table.

    (a) On crossing the third period from left to right there is a general increase in the first ionisation energy of the elements.

        (i) Why does the first ionisation energy increase across the period?

        1

        (ii) Write an equation corresponding to the first ionisation energy of chlorine.

        1

    (b) The electronegativities of elements in the third period are listed on page 10 of the databook.

        Why is no value provided for the noble gas, argon?

        1
        (3)

[Turn over

DO NOT
WRITE IN
THIS
MARGIN

Marks

3. A student writes the following two statements. **Both are incorrect**. In each case explain the mistake in the student's reasoning.

(a) All ionic compounds are solids at room temperature. Many covalent compounds are gases at room temperature. This proves that ionic bonds are stronger than covalent bonds.

1

(b) The formula for magnesium chloride is $MgCl_2$ because, in solid magnesium chloride, each magnesium ion is bonded to two chloride ions.

1

(2)

Marks

4. Petrol is a complex blend of many chemicals.

(*a*) A typical hydrocarbon found in petrol is shown below.

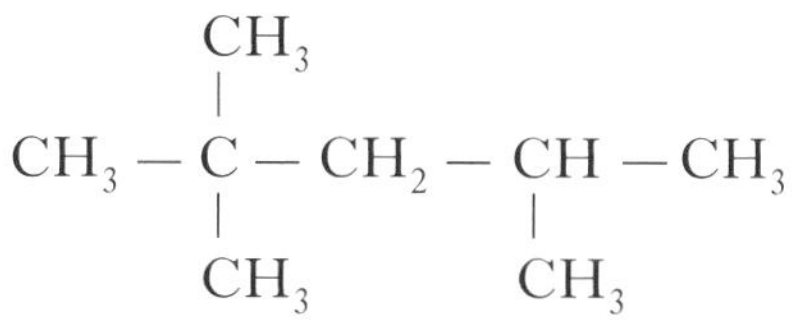

$$CH_3 - \underset{\underset{CH_3}{|}}{\overset{\overset{CH_3}{|}}{C}} - CH_2 - \underset{\underset{CH_3}{|}}{CH} - CH_3$$

What is the systematic name for this compound?

1

(*b*) In what way is a petrol that has been blended for use in winter different from a summer blend?

1

(*c*) The ester methyl stearate is also a useful vehicle fuel.

A student prepared this ester from methanol and stearic acid during the Prescribed Practical Activity, "Making Esters".

Describe how this ester was prepared.

2

(4)

DO NOT WRITE IN THIS MARGIN

Marks

5. Chlorine gas can be produced by heating calcium hypochlorite, $Ca(OCl)_2$, in dilute hydrochloric acid.

$$Ca(OCl)_2(s) + 2HCl(aq) \rightarrow Ca(OH)_2(aq) + 2Cl_2(g)$$

(*a*) Calculate the mass of calcium hypochlorite that would be needed to produce 0·096 litres of chlorine gas.

(Take the molar volume of chlorine gas to be 24 litres mol^{-1}.)

Show your working clearly. **2**

Marks

5. (continued)

(b) Chlorine is used in the manufacture of herbicides such as
2,4-dichlorophenoxyethanoic acid.

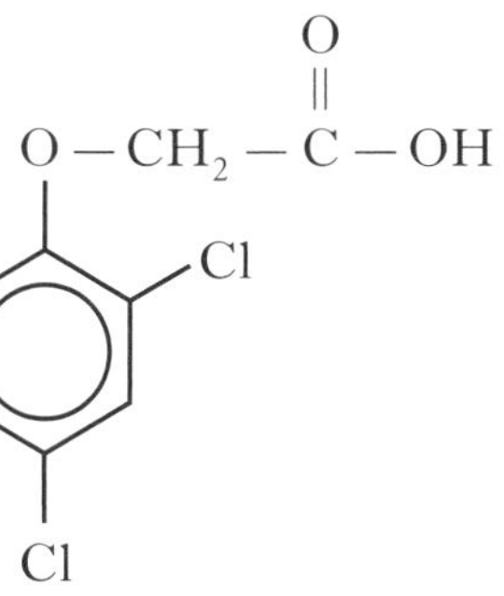

$$O-CH_2-\underset{\underset{O}{\parallel}}{C}-OH$$

(with benzene ring bearing Cl at positions 2 and 4)

Another commonly used herbicide is 4-chloro-2-methylphenoxyethanoic acid.

Draw a structural formula for 4-chloro-2-methylphenoxyethanoic acid.

1

(3)

[Turn over

Marks

6. Hairspray is a mixture of chemicals.

(*a*) A primary alcohol, 2-methylpropan-1-ol, is added to hairspray to help it dry quickly on the hair.

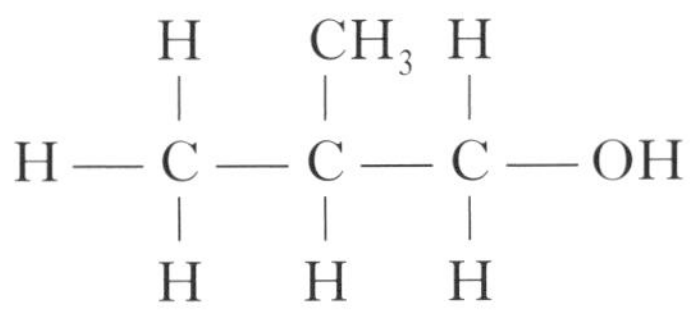

$$
\begin{array}{ccccc}
& H & & CH_3 & H \\
& | & & | & | \\
H\!-\!\!&C&\!-\!\!&C&\!-\!\!C\!-\!OH \\
& | & & | & | \\
& H & & H & H
\end{array}
$$

Draw a structural formula for a secondary alcohol that is an isomer of 2-methylpropan-1-ol.

1

Marks

(b) Triethanol amine and triisopropyl amine are bases used to neutralise acidic compounds in the hairspray to prevent damage to the hair.

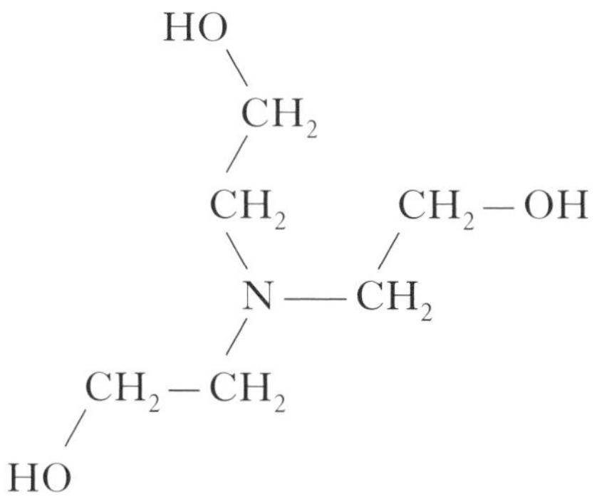

$$HO$$
$$\diagdown$$
$$CH_2$$
$$\diagup$$
$$CH_2 \qquad CH_2-OH$$
$$\diagdown \quad \diagup$$
$$N-CH_2$$
$$\diagup$$
$$CH_2-CH_2$$
$$\diagdown$$
$$HO$$

triethanol amine

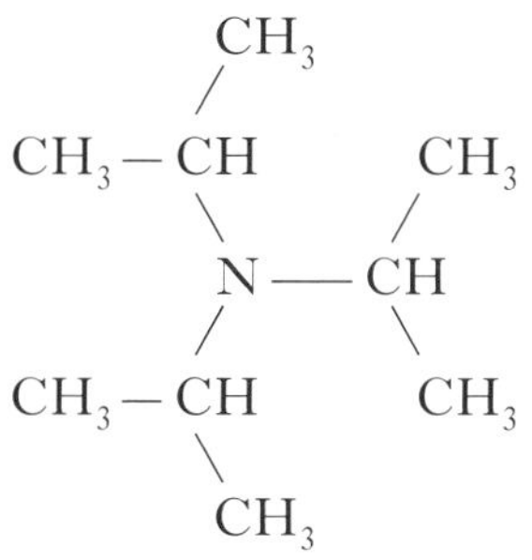

$$CH_3$$
$$\diagup$$
$$CH_3-CH \qquad CH_3$$
$$\diagdown \qquad \diagup$$
$$N-CH$$
$$\diagup \qquad \diagdown$$
$$CH_3-CH \qquad CH_3$$
$$\diagdown$$
$$CH_3$$

triisopropyl amine

molecular mass   149

boiling point   335 °C

molecular mass   143

boiling point   47 °C

In terms of the intermolecular bonding present, **explain clearly** why triethanol amine has a much higher boiling point than triisopropyl amine.

2

(3)

[**Turn over**

Marks

7. Paracetamol is a widely used painkiller.

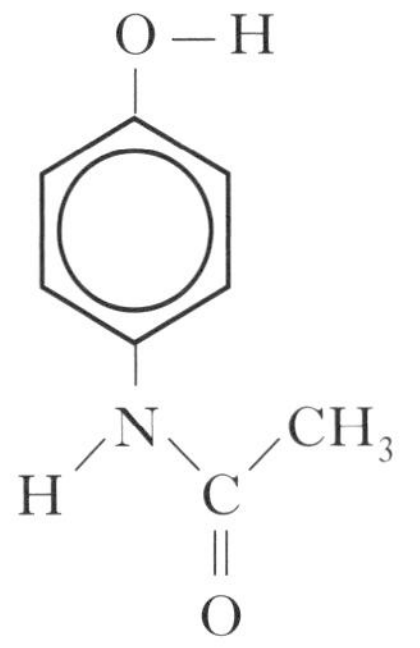

$$O-H$$

$$N \quad CH_3$$
$$H \quad C$$
$$\| $$
$$O$$

(*a*) Write the molecular formula for paracetamol.

1

(*b*) One antidote for paracetamol overdose is methionine.

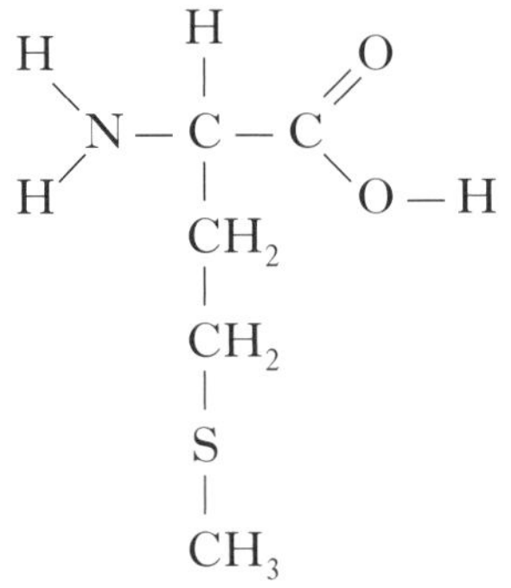

$$H \quad O$$
$$N-C-C$$
$$H \quad O-H$$
$$CH_2$$
$$CH_2$$
$$S$$
$$CH_3$$

To what family of organic compounds does methionine belong?

1

Marks

7.   **(continued)**

(c)   The concentration of paracetamol in a solution can be determined by measuring how much UV radiation it absorbs.

The graph shows how the absorbance of a sample containing $0.040\,g\,l^{-1}$ paracetamol varies with wavelength.

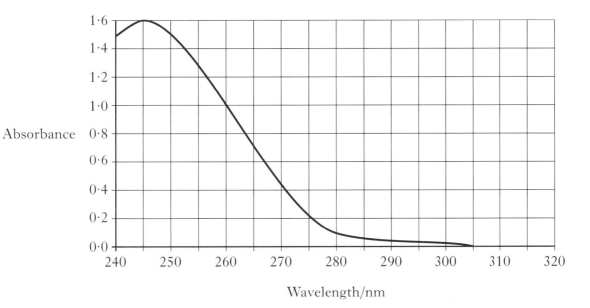

The quantity of UV radiation of wavelength 245 nm absorbed is directly proportional to the concentration of paracetamol.

The absorbance of a second sample of paracetamol solution measured at 245 nm was 0.90.

What is the concentration, in $g\,l^{-1}$, of this second paracetamol solution?

1

(3)

[Turn over

Marks

8. Diols are widely used in the manufacture of polyester polymers.

Polyethylene naphthalate is used to manufacture food containers. The monomers used to produce this polymer are shown.

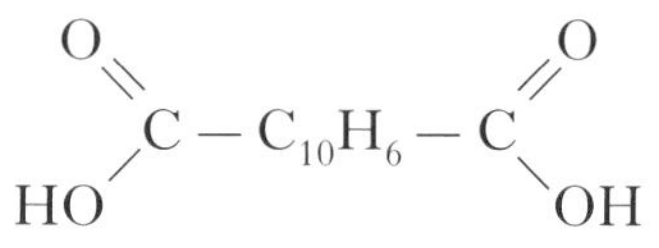

$$O \diagdown$$
$$C - C_{10}H_6 - C$$
$$HO \diagup \qquad\qquad \diagdown OH$$
naphthalenedicarboxylic acid

$$HO - CH_2 - CH_2 - OH$$

ethane-1,2-diol

(a) Draw the repeating unit for polyethylene naphthalate.

1

(b) Ethane-1,2-diol is produced in industry by reacting glycerol with hydrogen.

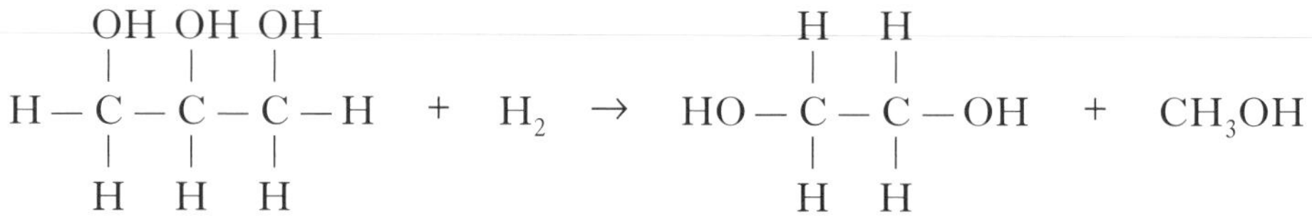

$$
\begin{array}{c}
\text{OH OH OH} \\
| \quad | \quad | \\
H - C - C - C - H \\
| \quad | \quad | \\
H \quad H \quad H
\end{array}
\; + \; H_2 \; \rightarrow \;
\begin{array}{c}
H \quad H \\
| \quad | \\
HO - C - C - OH \\
| \quad | \\
H \quad H
\end{array}
\; + \; CH_3OH
$$

glycerol   ethane-1,2-diol

Excess hydrogen reacts with 27·6 kg of glycerol to produce 13·4 kg of ethane-1,2-diol.

Calculate the percentage yield of ethane-1,2-diol.

Show your working clearly.

2

(3)

DO NOT WRITE IN THIS MARGIN

Marks

9. When vegetable oils are hydrolysed, mixtures of fatty acids are obtained. The fatty acids can be classified by their degree of unsaturation.

The table below shows the composition of each of the mixtures of fatty acids obtained when palm oil and olive oil were hydrolysed.

	Palm oil	**Olive oil**
Saturated fatty acids	51%	16%
Monounsaturated fatty acids	39%	75%
Polyunsaturated fatty acids	10%	9%

(*a*) Why does palm oil have a higher melting point than olive oil? **1**

(*b*) One of the fatty acids produced by the hydrolysis of palm oil is linoleic acid, $C_{17}H_{31}COOH$.

To which class (saturated, monounsaturated or polyunsaturated) does this fatty acid belong? **1**

(*c*) When a mixture of palm oil and olive oil is hydrolysed using a solution of sodium hydroxide, a mixture of the sodium salts of the fatty acids is obtained.

State a use for these fatty acid salts. **1**

(3)

[Turn over

Marks

10. Christian Schoenbein discovered ozone, O_3, in 1839.

(a) Ozone in air can be detected using paper strips that have been soaked in a mixture of starch and potassium iodide solution. The paper changes colour when ozone is present.

Ozone reacts with potassium iodide and water to form iodine, oxygen and potassium hydroxide.

(i) Write the balanced chemical equation for this reaction.

1

(ii) What colour would be seen on the paper when ozone is present?

1

(b) Ozone and oxygen gases are produced at the same electrode during the electrolysis of dilute sulphuric acid.

The ion-electron equation for the production of ozone is:

$$3H_2O(\ell) \rightarrow O_3(g) + 6H^+(aq) + 6e^-$$

Draw a labelled diagram of the assembled apparatus that could be used to carry out the electrolysis of dilute sulphuric acid, showing how the ozone/oxygen gas mixture can be collected.

2

DO NOT WRITE IN THIS MARGIN

Marks

10. (continued)

(*c*) When ozone is bubbled through a solution containing an alkene, an ozonolysis reaction takes place.

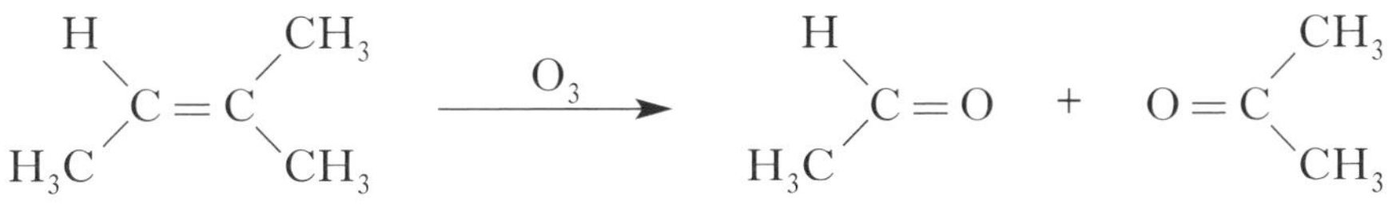

compound **X** compound **Y**

(i) 2 cm^3 of an oxidising agent was added to 5 cm^3 of compound **X** in a test tube. After a few minutes a colour change from orange to green was observed.

Name the oxidising agent used. **1**

(ii) Draw a structural formula for the alkene which, on ozonolysis, would produce propanal and butan-2-one. **1**

(6)

Marks

11. Sulphurous acid, H_2SO_3, is a weak acid produced when sulphur-containing compounds in fuels are burned.

   (a) What is meant by a **weak** acid?

1

   (b) The table below shows the results of two experiments that were carried out to compare sulphurous acid with the strong acid, hydrochloric acid.

	Sulphurous acid $0.1\ mol\,l^{-1}$	Hydrochloric acid $0.1\ mol\,l^{-1}$
Rate of reaction with strip of magnesium	slow	fast
Volume of acid required to neutralise $20\ cm^3$ of $0.1\ mol\,l^{-1}$ sodium hydroxide solution	$10\ cm^3$	$20\ cm^3$

   (i) Why did the magnesium react more quickly with the hydrochloric acid?

1

   (ii) Why is a smaller volume of sulphurous acid solution needed to neutralise the sodium hydroxide solution?

1

DO NOT WRITE IN THIS MARGIN

Marks

11. (*b*) (continued)

(iii) The concentration of the sodium hydroxide solution used was $0{\cdot}1\ mol\,l^{-1}$.

Calculate the pH of this solution. **1**

(4)

[Turn over

Marks

12. The element iodine has only one isotope that is stable. Several of the radioactive isotopes of iodine have medical uses. Iodine-131, for example, is used in the study of the thyroid gland and it decays by beta emission.

(*a*) Why are some atoms unstable?

1

(*b*) Complete the balanced nuclear equation for the beta decay of iodine-131.

$$^{131}_{53}I \quad \rightarrow$$

1

(*c*) The graph shows how the mass of iodine-131 in a sample changes over a period of time.

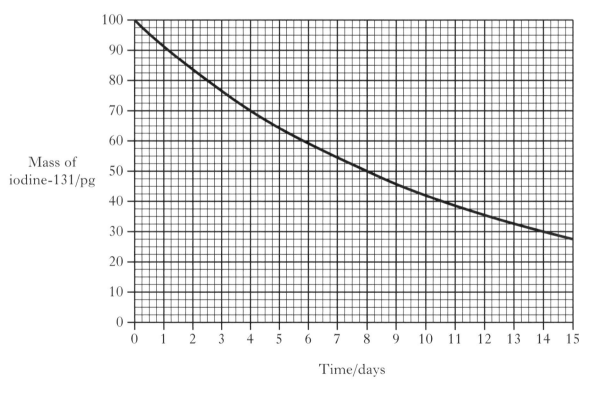

Mass of iodine-131/pg

Time/days

(i) What is the half-life of this isotope?

1

DO NOT WRITE IN THIS MARGIN

Marks

12. (*c*) (continued)

(ii) A sample of sodium iodide solution contained 100 pg of iodine-131 when it was prepared.

Four days later it was injected into a patient.

How many $^{131}I^{-}$ ions would the 4 day old sample contain?

($1\ pg = 1 \times 10^{-12}\ g$)

2

(5)

[Turn over

DO NOT WRITE IN THIS MARGIN

Marks

13. Rivers and drains are carefully monitored to ensure that they remain uncontaminated by potentially harmful substances from nearby industries. Chromate ions, CrO_4^{2-}, are particularly hazardous.

(*a*) When chromate ions dissolve in water the following equilibrium is established.

$$\underset{\text{yellow}}{2CrO_4^{2-}(aq)} + 2H^+(aq) \rightleftharpoons \underset{\text{orange}}{Cr_2O_7^{2-}(aq)} + H_2O(\ell)$$

Explain fully the colour change that would be observed when solid sodium hydroxide is added to the solution. **2**

(*b*) The concentration of chromate ions in water can be measured by titrating with a solution of iron(II) sulphate solution.

(i) To prepare the iron(II) sulphate solution used in this titration, iron(II) sulphate crystals were weighed accurately into a dry beaker.

Describe how these crystals should be dissolved and then transferred to a standard flask in order to produce a solution of accurately known concentration. **2**

Marks

13. (*b*) **(continued)**

    (ii) A $50 \cdot 0\,cm^3$ sample of contaminated water containing chromate ions was titrated and found to require $27 \cdot 4\,cm^3$ of $0 \cdot 0200\,mol\,l^{-1}$ iron(II) sulphate solution to reach the end-point.

    The redox equation for the reaction is:

$$3Fe^{2+}(aq)\ +\ CrO_4^{2-}(aq)\ +\ 8H^+(aq)\ \rightarrow\ 3Fe^{3+}(aq)\ +\ Cr^{3+}(aq)\ +\ 4H_2O(\ell)$$

    Calculate the chromate ion concentration, in $mol\,l^{-1}$, present in the sample of water.

    Show your working clearly.

2

(6)

[Turn over

Marks

14. The enthalpies of combustion of some alcohols are shown in the table.

Name of alcohol	Enthalpy of combustion/kJ mol^{-1}
methanol	–727
ethanol	–1367
propan-1-ol	–2020

(a) Using this data, predict the enthalpy of combustion of butan-1-ol, in kJ mol^{-1}.

1

(b) A value for the enthalpy of combustion of butan-2-ol, C_4H_9OH, can be determined experimentally using the apparatus shown.

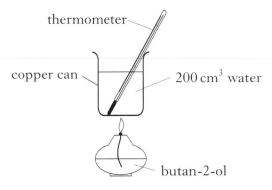

Mass of butan-2-ol burned   =   1·0 g
Temperature rise of water   =   40 °C

Use these results to calculate the enthalpy of combustion of butan-2-ol, in kJ mol^{-1}.

2

DO NOT WRITE IN THIS MARGIN

Marks

14. (continued)

(*c*) Enthalpy changes can also be calculated using Hess's Law.

The enthalpy of formation for pentan-1-ol is shown below.

$$5C(s) \ + \ 6H_2(g) \ + \ \frac{1}{2}O_2(g) \ \rightarrow \ C_5H_{11}OH(\ell) \qquad \Delta H = -354\,kJ\ mol^{-1}$$

Using this value, and the enthalpies of combustion of carbon and hydrogen from the data booklet, calculate the enthalpy of combustion of pentan-1-ol, in $kJ\ mol^{-1}$.

2

(5)

[Turn over

15.   Cerium metal is extracted from the mineral monazite.

   The flow diagram for the extraction of cerium from the mineral is shown below.

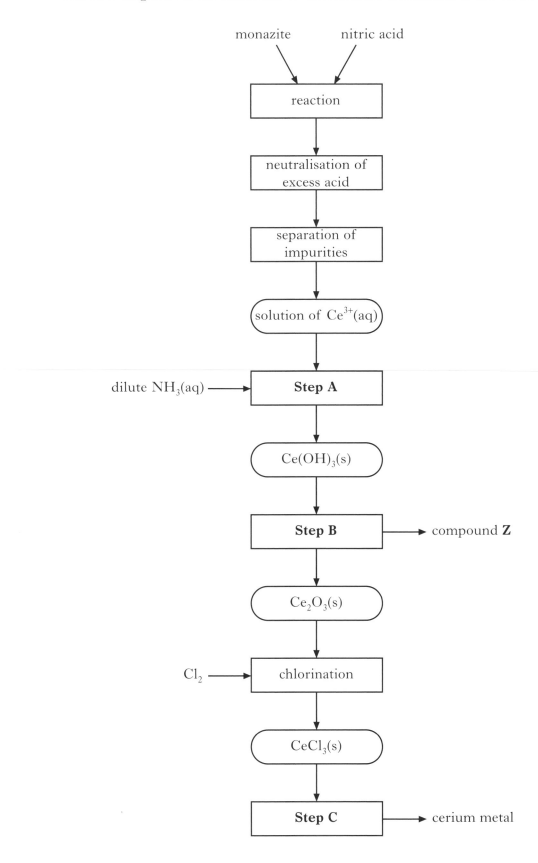

DO NOT WRITE IN THIS MARGIN

Marks

15. (continued)

(*a*) Name the type of chemical reaction taking place in **Step A**. 1

(*b*) In **Step B**, cerium hydroxide is heated to form cerium oxide, Ce_2O_3, and compound **Z**.

Name compound **Z**. 1

(*c*) In **Step C**, cerium metal is obtained by electrolysis.

(i) What feature of the electrolysis can be used to reduce the cost of cerium production? 1

(ii) The equation for the reaction at the negative electrode is

$$Ce^{3+} + 3e^- \rightarrow Ce$$

Calculate the mass of cerium, in kg, produced in 10 minutes when a current of 4000 A is used.

Show your working clearly. 2

(5)

[Turn over for Question 16 on *Page thirty-four*

Marks

16. The boiling point of water can be raised by the addition of a solute.

The increase in boiling point depends only on the **number** of solute particles but not the type of particle.

The increase in boiling point (ΔT_b), in °C, can be estimated using the formula shown.

$$\Delta T_b = 0 \cdot 51 \times c \times i$$

where

c is the concentration of the solution in mol l^{-1}.

i is the number of particles released into solution when one formula unit of the solute dissolves.

The value of i for a number of compounds is shown in the table below.

Solute	i
NaCl	2
$MgCl_2$	3
$(NH_4)_3PO_4$	4

(a) What is the value of i for sodium sulphate?

1

(b) Calculate the increase in boiling point, ΔT_b, for a 0·10 mol l^{-1} solution of ammonium phosphate.

1

(2)

[END OF QUESTION PAPER]

ADDITIONAL GRAPH PAPER FOR USE IN QUESTION 1(*b*) (ii)

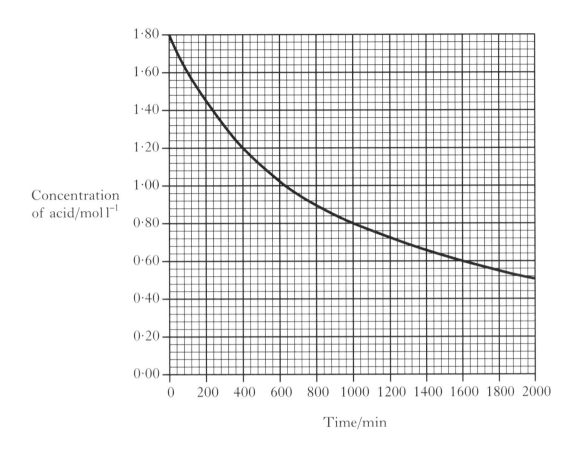

ADDITIONAL SPACE FOR ANSWERS

HIGHER

2012

[BLANK PAGE]

FOR OFFICIAL USE

Total
Section B

X012/12/02

NATIONAL
QUALIFICATIONS
2012

MONDAY, 14 MAY
1.00 PM – 3.30 PM

CHEMISTRY
HIGHER

Fill in these boxes and read what is printed below.

Full name of centre

Town

Forename(s)

Surname

Date of birth

Day   Month   Year   Scottish candidate number   Number of seat

Reference may be made to the Chemistry Higher and Advanced Higher Data Booklet.

SECTION A—Questions 1–40 (40 marks)

Instructions for completion of **Section A** are given on page two.

For this section of the examination you must use an **HB pencil**.

SECTION B (60 marks)

1   All questions should be attempted.

2   The questions may be answered in any order but all answers are to be written in the spaces provided in this answer book, **and must be written clearly and legibly in ink**.

3   Rough work, if any should be necessary, should be written in this book and then scored through when the fair copy has been written. If further space is required, a supplementary sheet for rough work may be obtained from the Invigilator.

4   Additional space for answers will be found at the end of the book. If further space is required, supplementary sheets may be obtained from the Invigilator and should be inserted inside the **front** cover of this book.

5   The size of the space provided for an answer should not be taken as an indication of how much to write. It is not necessary to use all the space.

6   Before leaving the examination room you must give this book to the Invigilator. If you do not, you may lose all the marks for this paper.

SECTION A

Read carefully

1. Check that the answer sheet provided is for **Chemistry Higher (Section A)**.

2. For this section of the examination you must use an **HB pencil** and, where necessary, an eraser.

3. Check that the answer sheet you have been given has **your name**, **date of birth**, **SCN** (Scottish Candidate Number) and **Centre Name** printed on it.

   Do not change any of these details.

4. If any of this information is wrong, tell the Invigilator immediately.

5. If this information is correct, **print** your name and seat number in the boxes provided.

6. The answer to each question is **either** A, B, C or D. Decide what your answer is, then, using your pencil, put a horizontal line in the space provided (see sample question below).

7. There is only **one correct answer** to each question.

8. Any rough working should be done on the question paper or the rough working sheet, **not** on your answer sheet.

9. At the end of the examination, put the **answer sheet for Section A inside the front cover of your answer book**.

Sample Question

To show that the ink in a ball-pen consists of a mixture of dyes, the method of separation would be

    A chromatography

    B fractional distillation

    C fractional crystallisation

    D filtration.

The correct answer is **A**—chromatography. The answer **A** has been clearly marked in **pencil** with a horizontal line (see below).

Changing an answer

If you decide to change your answer, carefully erase your first answer and using your pencil, fill in the answer you want. The answer below has been changed to **D**.

1. Isotopes of an element have

   A   the same mass number

   B   the same number of neutrons

   C   equal numbers of protons and neutrons

   D   different numbers of neutrons.

2. Four metals **W**, **X**, **Y** and **Z** and their compounds behaved as described.

   (i)   Only **X**, **Y** and **Z** reacted with dilute hydrochloric acid.

   (ii) The oxides of **W**, **X** and **Y** were reduced to the metal when heated with carbon powder. The oxide of **Z** did not react.

   (iii) A displacement reaction occurred when **X** was added to an aqueous solution of the nitrate of **Y**.

   What is the correct order of reactivity of these metals (most reactive first)?

   A   W, Y, X, Z

   B   W, X, Y, Z

   C   Z, X, Y, W

   D   Z, Y, X, W

3. A positively charged particle with electron arrangement 2, 8 could be

   A   a neon atom

   B   a fluoride ion

   C   a sodium atom

   D   an aluminium ion.

4. A solution of potassium carbonate, made up using tap water, was found to be cloudy.

   This could result from the tap water containing

   A   sodium ions

   B   chloride ions

   C   magnesium ions

   D   sulphate ions.

   (You may wish to refer to the Data Booklet.)

5. 1 mol of hydrogen gas and 1 mol of iodine vapour were mixed and allowed to react. After t seconds, 0·8 mol of hydrogen remained.

   The number of moles of hydrogen iodide formed at t seconds was

   A   0·2

   B   0·4

   C   0·8

   D   1·6.

6. In a reaction involving gases, an increase in temperature results in

   A   an increase in activation energy

   B   an increase in the enthalpy change

   C   a decrease in the activation energy

   D   more molecules per second forming an activated complex.

7. Calcium carbonate reacts with nitric acid as follows.

   $CaCO_3(s) + 2HNO_3(aq) \rightarrow Ca(NO_3)_2(aq) + H_2O(\ell) + CO_2(g)$

   0·05 mol of calcium carbonate was added to a solution containing 0·08 mol of nitric acid.

   Which of the following statements is true?

   A   0·05 mol of carbon dioxide is produced.

   B   0·08 mol of calcium nitrate is produced.

   C   Calcium carbonate is in excess by 0·01 mol.

   D   Nitric acid is in excess by 0·03 mol.

   [Turn over

8.

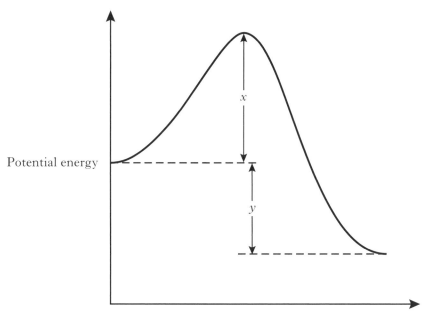

Reaction pathway

The enthalpy change for the forward reaction can be represented by

A x

B y

C $x + y$

D $x - y$.

9. $5N_2O_4(\ell) + 4CH_3NHNH_2(\ell) \rightarrow 4CO_2(g) + 12H_2O(\ell) + 9N_2(g)$ $\Delta H = -5116\,kJ$

The energy released when 2 moles of each reactant are mixed and ignited is

A 2046 kJ

B 2558 kJ

C 4093 kJ

D 5116 kJ.

10. Atoms of nitrogen and element **X** form a bond in which the electrons are shared equally.

Element **X** could be

A carbon

B oxygen

C chlorine

D phosphorus.

11. Which line in the table represents the solid in which only van der Waals' forces are overcome when the substance melts?

	Melting point / °C	Electrical conduction of solid
A	714	non-conductor
B	98	conductor
C	660	conductor
D	44	non-conductor

12. Which of the following does **not** contain covalent bonds?

    A Hydrogen gas

    B Helium gas

    C Nitrogen gas

    D Solid sulphur

13. Which of the following structures is **never** found in compounds?

    A Ionic

    B Monatomic

    C Covalent network

    D Covalent molecular

14. In which of the following solvents is lithium chloride most likely to dissolve?

    A Hexane

    B Benzene

    C Methanol

    D Tetrachloromethane

15. A balloon contains 0·1 mol of oxygen gas, and 0·2 mol of carbon dioxide gas.

    The total number of molecules in the balloon is approximately

    A $6·0 \times 10^{23}$

    B $3·6 \times 10^{23}$

    C $2·4 \times 10^{23}$

    D $1·8 \times 10^{23}$.

16. Which of the following pairs of gases occupy the same volume?

    (Assume all measurements are made under the same conditions of temperature and pressure.)

    A 2 g hydrogen and 14 g nitrogen

    B 32 g methane and 88 g carbon dioxide

    C 7 g carbon monoxide and 16 g oxygen

    D 10 g hydrogen chloride and 10 g sulphur dioxide

17. $2NO(g) + O_2(g) \rightarrow 2NO_2(g)$

    How many litres of nitrogen dioxide gas could theoretically be obtained in the reaction of 1 litre of nitrogen monoxide gas with 2 litres of oxygen gas?

    (All volumes are measured under the same conditions of temperature and pressure.)

    A 1

    B 2

    C 3

    D 4

18. Which of the following hydrocarbons is **least** likely to be added to petrol to improve the efficiency of its burning?

    A $CH_3CH_2CH_2CH_2CH_2CH_2CH_3$

    B
$$CH_3CHCHCHCH_3$$
    with CH_3, CH_3, CH_3 substituents

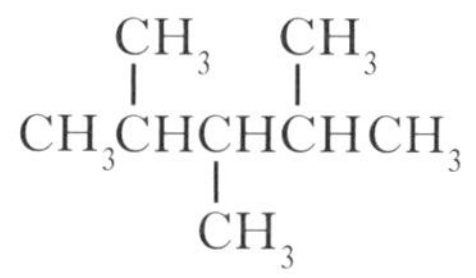

    C
a cyclohexane ring drawn with CH_2, H_2C, CH_2, H_2C, CH_2, CH_2

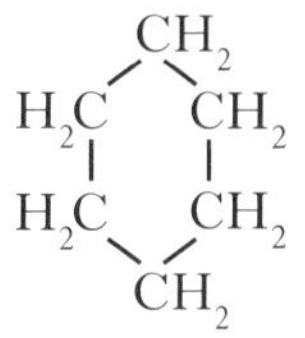

    D

19. Biogas is produced under anaerobic conditions by the fermentation of biological materials.

   What is the main constituent of biogas?

   A   Butane

   B   Ethane

   C   Methane

   D   Propane

20. Which equation represents a reaction which takes place during reforming?

   A $C_6H_{14} \rightarrow C_6H_6 + 4H_2$

   B $C_4H_8 + H_2 \rightarrow C_4H_{10}$

   C $C_2H_5OH \rightarrow C_2H_4 + H_2O$

   D $C_8H_{18} \rightarrow C_4H_{10} + C_4H_8$

21. Which of the following is an isomer of 2-methyl pentane?

   A
   $$\begin{array}{cc} CH_3 & CH_3 \\ | & | \\ CH - & CH \\ | & | \\ CH_3 & CH_3 \end{array}$$

   B
   $$\begin{array}{cc} CH_3 & \\ | & \\ CH - & CH_2 \\ | & | \\ CH_3 & CH_2 - CH_3 \end{array}$$

   C $CH_3 - CH_2 - CH_2 - \underset{\underset{CH_3}{|}}{\overset{\overset{CH_3}{|}}{CH}}$

   D $CH_3 - \underset{\underset{}{\overset{\overset{CH_3}{|}}{CH}}}{} - CH_2 - \underset{}{\overset{\overset{CH_3}{|}}{CH_2}}$

22. Which line in the table shows the correct functional group for each homologous series?

	Alkanoic acid	Alkanol	Alkanal
A	$-C{\overset{\displaystyle O}{\diagdown}}_H$	$-OH$	$-C{\overset{\displaystyle O}{\diagdown}}_{OH}$
B	$-C{\overset{\displaystyle O}{\diagdown}}_{OH}$	$-OH$	$-C{\overset{\displaystyle O}{\diagdown}}_H$
C	$-C{\overset{\displaystyle O}{\diagdown}}_{OH}$	$-C{\overset{\displaystyle O}{\diagdown}}_H$	$-OH$
D	$-OH$	$-C{\overset{\displaystyle O}{\diagdown}}_{OH}$	$-C{\overset{\displaystyle O}{\diagdown}}_H$

23. Hydrolysis of an ester gave an alkanol and an alkanoic acid both of which had the same molecular mass of 60.

   The structure of the ester was

   A
   $$H - \underset{\underset{H}{|}}{\overset{\overset{H}{|}}{C}} - \overset{\overset{O}{\|}}{C} - O - \underset{\underset{H}{|}}{\overset{\overset{H}{|}}{C}} - \underset{\underset{H}{|}}{\overset{\overset{H}{|}}{C}} - H$$

   B
   $$H - \underset{\underset{H}{|}}{\overset{\overset{H}{|}}{C}} - \overset{\overset{O}{\|}}{C} - O - \underset{\underset{H}{|}}{\overset{\overset{H}{|}}{C}} - \underset{\underset{H}{|}}{\overset{\overset{H}{|}}{C}} - \underset{\underset{H}{|}}{\overset{\overset{H}{|}}{C}} - H$$

   C
   $$H - \underset{\underset{H}{|}}{\overset{\overset{H}{|}}{C}} - \underset{\underset{H}{|}}{\overset{\overset{H}{|}}{C}} - \overset{\overset{O}{\|}}{C} - O - \underset{\underset{H}{|}}{\overset{\overset{H}{|}}{C}} - H$$

   D
   $$H - \underset{\underset{H}{|}}{\overset{\overset{H}{|}}{C}} - \underset{\underset{H}{|}}{\overset{\overset{H}{|}}{C}} - \overset{\overset{O}{\|}}{C} - O - \underset{\underset{H}{|}}{\overset{\overset{H}{|}}{C}} - \underset{\underset{H}{|}}{\overset{\overset{H}{|}}{C}} - H$$

24. Which of the following statements about benzene is true?

    (You may wish to refer to the Data Booklet.)

    A Benzene has the same ratio of carbon to hydrogen as ethyne.

    B Benzene reacts with copper(II) oxide more easily than ethanol.

    C Benzene is more volatile than ethanal.

    D Benzene undergoes addition reactions more readily than ethene.

25. Which of the following is a possible product of the reaction of propyne with bromine?

A
```
  Br    H H
    \   | |
     C = C — C — H
    /       |
  H         H
```

B
```
    Br  Br  H
    |   |   |
Br— C — C — C — H
    |   |   |
    Br  H   H
```

C
```
  Br    Br  H
    \   |   |
     C = C — C — H
    /       |
  H         H
```

D
```
    Br  H   Br
    |   |   |
 H— C — C — C — H
    |   |   |
    Br  H   Br
```

26. Which alcohol could be oxidised to a carboxylic acid?

A
```
  H  H  H  H  H
  |  |  |  |  |
H—C—C—C—C—C—H
  |  |  |  |  |
  H  H  H  OH H
```

B
```
  H  H  H  H  H
  |  |  |  |  |
H—C—C—C—C—C—H
  |  |  |  |  |
  H  H  OH H  H
```

C
```
              H
              |
          H—C—H
    H     |   H  H
    |     |   |  |
 H—C—C—C—C—H
    |  |  |  |
    H  OH H  H
```

D
```
          H
          |
      H—C—H
    H     |   H
    |     |   |
 H—C—C—C—OH
    |  |  |
    H  |  H
    H—C—H
       |
       H
```

27. What mixture of gases is known as synthesis gas?

    A Methane and oxygen

    B Carbon monoxide and oxygen

    C Carbon dioxide and hydrogen

    D Carbon monoxide and hydrogen

28. Which of the following substances does **not** have delocalised electrons?

    A Aluminium

    B Poly(ethyne)

    C Poly(ethenol)

    D Carbon (graphite)

[Turn over

29. The arrangement of amino acids in a peptide is

$$Z-X-W-V-Y$$

where the letters V, W, X, Y and Z represent amino acids.

On partial hydrolysis of the peptide, which of the following sets of dipeptides is possible?

A   V-Y, Z-X, W-Y, X-W

B   Z-X, V-Y, W-V, X-W

C   Z-X, X-V, W-V, V-Y

D   X-W, X-Z, Z-W, Y-V

30. Aluminium reacts with oxygen to form aluminium oxide.

$$2Al(s) + 1\frac{1}{2}O_2(g) \rightarrow Al_2O_3(s) \quad \Delta H = -1670 \text{ kJ mol}^{-1}$$

What is the enthalpy of combustion of aluminium in kJ mol^{-1}?

A   −835

B   −1113

C   −1670

D   +1670

31. A few drops of concentrated sulphuric acid were added to a mixture of 0·1 mol of methanol and 0·2 mol of ethanoic acid. Even after a considerable time, the reaction mixture was found to contain some of each reactant.

Which of the following is the best explanation for the incomplete reaction?

A   The temperature was too low.

B   An equilibrium mixture was formed.

C   Insufficient methanol was used.

D   Insufficient ethanoic acid was used.

32. Which line in the table applies correctly to the use of a catalyst in a chemical reaction?

	Position of equilibrium	Effect on value of ΔH
A	Moved to right	Decreased
B	Unaffected	Increased
C	Moved to left	Unaffected
D	Unaffected	Unaffected

33. The hypochlorite ion, $ClO^-(aq)$, produced in the reaction shown, is used as a bleach.

$$Cl_2(g) + H_2O(\ell) \rightleftharpoons 2H^+(aq) + ClO^-(aq) + Cl^-(aq)$$

The concentration of ClO^- ions could be increased by the addition of

A   solid potassium hydroxide

B   concentrated hydrochloric acid solution

C   solid sodium chloride

D   solid potassium sulphate.

34. A solution of hydrochloric acid with a pH of 6 and another of sodium hydroxide with a pH of 8 are each diluted by a factor of 100.

After dilution, when tested using pH indicator paper,

A   the pH of the acid drops by 2

B   the pH of the alkali rises by 2

C   the pH of the acid equals that of the alkali

D   the pH of the acid is 2 below that of the alkali.

35. Equal volumes of four 1 mol l^{-1} solutions were compared.

Which of the following 1 mol l^{-1} solutions contains the fewest ions?

A   Hydrochloric acid

B   Ethanoic acid

C   Sodium chloride

D   Sodium hydroxide

36. Equal volumes of 0.1 mol l^{-1} solutions of sodium hydroxide and propanoic acid were mixed together.

    The pH of the resulting solution could be

    A   3

    B   5

    C   7

    D   9.

37. During a redox process in acid solution, iodate ions, $IO_3^-(aq)$, are converted into iodine, $I_2(aq)$.

$$IO_3^-(aq) \rightarrow I_2(aq)$$

    The numbers of $H^+(aq)$ and $H_2O(\ell)$ required to balance the ion-electron equation for the formation of 1 mol of $I_2(aq)$ are, respectively

    A   3 and 6

    B   6 and 3

    C   6 and 12

    D   12 and 6.

38. One mole of metal would be deposited by passing $96\,500$ C through a solution of

    A   silver(I) nitrate

    B   gold(III) nitrate

    C   nickel(II) nitrate

    D   copper(II) nitrate.

39. $^2_1H + {}^3_1H \rightarrow {}^4_2He + {}^1_0n$

    The above process represents

    A   nuclear fission

    B   nuclear fusion

    C   proton capture

    D   neutron capture.

40. Naturally occurring nitrogen consists of two isotopes ^{14}N and ^{15}N.

    How many different nitrogen molecules will occur in the air?

    A   1

    B   2

    C   3

    D   4

Candidates are reminded that the answer sheet MUST be returned INSIDE the front cover of this answer book.

[Turn over

[BLANK PAGE]

Marks

SECTION B

All answers must be written clearly and legibly in ink.

1. The elements lithium to neon make up the second period of the Periodic Table.

Li	Be	B	C	N	O	F	Ne

   (*a*) Name an element from the second period that exists as a covalent network.

1

   (*b*) Why do the atoms decrease in size from lithium to neon?

1

(2)

[Turn over

Marks

2. Copper(II) carbonate reacts with dilute hydrochloric acid as shown.

$$CuCO_3(s) + 2HCl(aq) \rightarrow CuCl_2(aq) + H_2O(\ell) + CO_2(g)$$

A student used the apparatus shown below to follow the progress of the reaction.

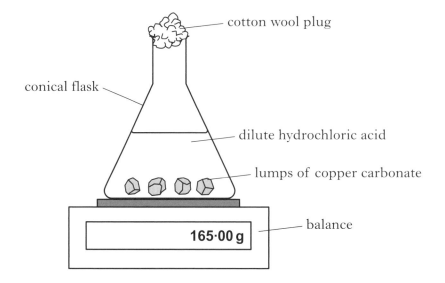

(*a*) Suggest why a cotton wool plug is placed in the mouth of the conical flask.

1

Marks

2. **(continued)**

   (*b*) The experiment was carried out using 0·50 g samples of both pure and impure copper(II) carbonate. The graph below shows the results obtained.

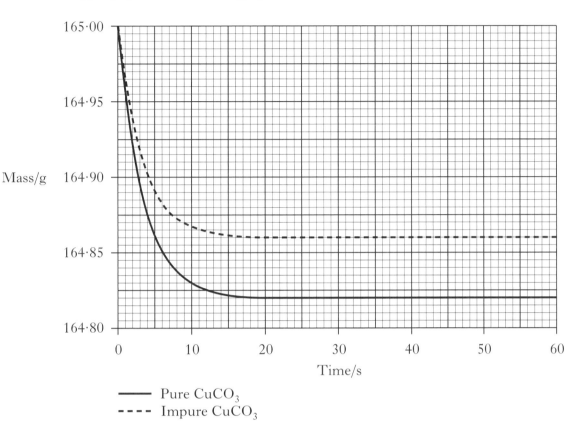

       —— Pure $CuCO_3$
       ---- Impure $CuCO_3$

   (i) For the sample of pure copper(II) carbonate, calculate the average reaction rate, in $g\,s^{-1}$, over the first 10 seconds.

1

   (ii) Calculate the mass, in grams, of copper(II) carbonate present in the impure sample.

   Show your working clearly.

1

(3)

Marks

3.   Ethanol, C_2H_5OH, can be used as a fuel in some camping stoves.

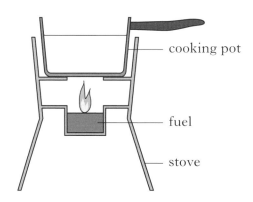

cooking pot

fuel

stove

(a)   The enthalpy of combustion of ethanol is $-1367\,kJ\,mol^{-1}$.

Using this value, calculate the number of moles of ethanol required to raise the temperature of 500 g of water from 18 °C to 100 °C.

Show your working clearly.

2

(b)   Suggest **two** reasons why less energy is obtained from burning ethanol in the camping stove than is predicted from its enthalpy of combustion.

2

(4)

DO NOT WRITE IN THIS MARGIN

Marks

4. Phosphorus-32 and strontium-89 are two radioisotopes used to study how far mosquitoes travel.

(*a*) Strontium-89 decays by emission of a beta particle.

Complete the nuclear equation for the decay of strontium-89.

$$^{89}Sr \rightarrow$$ **1**

(*b*) In an experiment, 10 g of strontium-89 chloride was added to a sugar solution used to feed mosquitoes.

(i) The strontium-89 chloride solution was fed to the mosquitoes in a laboratory at 20 °C. When the mosquitoes were released, the outdoor temperature was found to be 35 °C.

What effect would the increase in temperature have on the half-life of the strontium-89? **1**

(ii) Calculate the mass, in grams, of strontium-89 present in the 10 g sample of strontium-89 chloride, $SrCl_2$. **1**

(*c*) A mosquito fed on a solution containing phosphorus-32 is released.

Phosphorus-32 has a half-life of 14 days.

When the mosquito is recaptured 28 days later, what fraction of the phosphorus-32 will remain? **1**

(4)

5. The concentration of ethanol in a person's breath can be determined by measuring the voltage produced in an electrochemical cell.

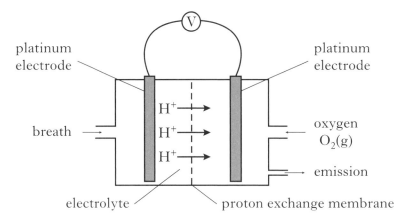

Different ethanol vapour concentrations produce different voltages as is shown in the graph below.

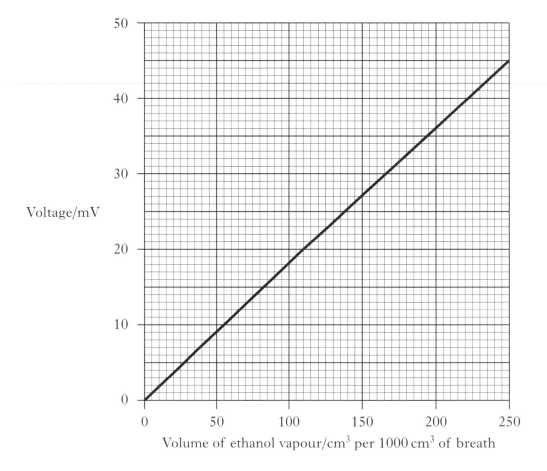

DO NOT WRITE IN THIS MARGIN

Marks

5. (continued)

(*a*) Calculate the number of ethanol molecules in 1000 cm^3 of breath when a voltage of 20 mV was recorded.

(Take the molar volume of ethanol vapour to be 24 litres mol^{-1}.)

Show your working clearly.

2

(*b*) The ion-electron equations for the reduction and oxidation reactions occurring in the cell are shown below.

$$O_2 + 4H^+ + 4e^- \rightarrow 2H_2O$$

$$CH_3CH_2OH + H_2O \rightarrow CH_3COOH + 4H^+ + 4e^-$$

Write the overall redox equation for the reaction taking place.

1

(*c*) Platinum metal acts as a heterogeneous catalyst in this reaction.

What is meant by a **heterogeneous catalyst**?

1

(4)

Marks

6. Compounds containing sulphur occur widely in nature.

    (a) The compound dimethyldisulphide, $CH_3S_2CH_3$, is present in garlic and onions.

    Draw a full structural formula for this compound.

    1

    (b) Hydrogen sulphide, H_2S, formed by the decomposition of proteins, can cause an unpleasant odour in water supplies.

        (i) Chlorine, added to the water, removes the hydrogen sulphide.

            The equation for the reaction taking place is

            $$4Cl_2(aq) + H_2S(aq) + 4H_2O(\ell) \rightarrow SO_4^{2-}(aq) + 10H^+(aq) + 8Cl^-(aq)$$

            An average of $29 \cdot 4\,cm^3$ of $0 \cdot 010\,mol\,l^{-1}$ chlorine solution was required to react completely with a $50 \cdot 0\,cm^3$ sample of water.

            Calculate the hydrogen sulphide concentration, in $mol\,l^{-1}$, present in the water sample.

            Show your working clearly.

    2

Marks

6. (*b*) (continued)

   (ii) Liquid hydrogen sulphide has a boiling point of −60 °C.

   Explain clearly why hydrogen sulphide is a gas at room temperature. In your answer, you should name the intermolecular forces involved and indicate how they arise.

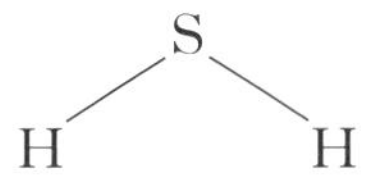

2

(5)

[Turn over

Marks

7. Polyurethanes are polymers that are widely used in industry. They are produced by the reaction of diisocyanates with diols.

   (a) The structure of one such diisocyanate is shown below.

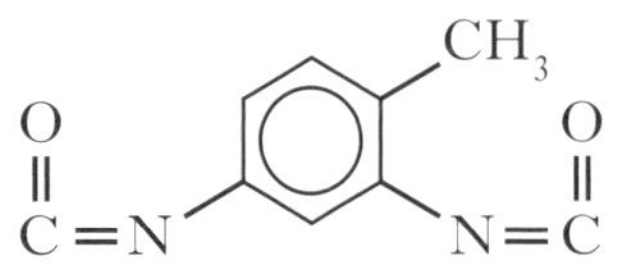

   The molecular formula for this compound can be written as $C_wH_xN_yO_z$.

   Give the values for **w**, **x**, **y** and **z**.

   w =   **x** =   **y** =   **z** =

   1

   (b) Lycra is a polyurethane polymer made by the polymerisation of 2-diisocyanatoethane with ethane-1,2-diol. The reaction to form the repeating unit is shown.

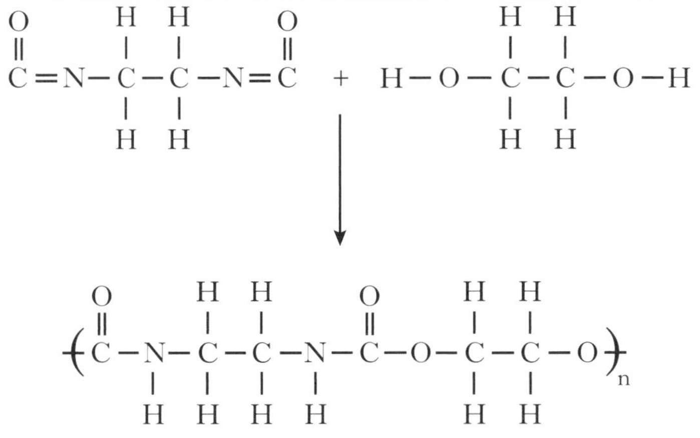

   Why is this polymerisation described as an addition reaction?

   1

Marks

7. **(continued)**

(*c*) Lycra is strong because of the hydrogen bonding between neighbouring polymer chains.

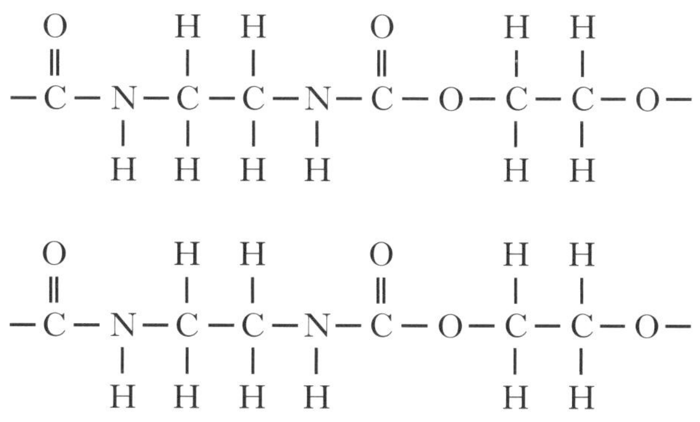

$$O \quad\quad H \quad H \quad\quad O \quad\quad H \quad H$$
$$\parallel \quad\quad | \quad | \quad\quad \parallel \quad\quad | \quad |$$
$$-C-N-C-C-N-C-O-C-C-O-$$
$$| \quad | \quad | \quad | \quad\quad\quad | \quad |$$
$$H \quad H \quad H \quad H \quad\quad\quad H \quad H$$

$$O \quad\quad H \quad H \quad\quad O \quad\quad H \quad H$$
$$\parallel \quad\quad | \quad | \quad\quad \parallel \quad\quad | \quad |$$
$$-C-N-C-C-N-C-O-C-C-O-$$
$$| \quad | \quad | \quad | \quad\quad\quad | \quad |$$
$$H \quad H \quad H \quad H \quad\quad\quad H \quad H$$

Draw a dotted line to show a hydrogen bond between the polymer chains above.

1

(3)

[Turn over

DO NOT WRITE IN THIS MARGIN

Marks

8. Aspartame is an artificial sweetener which has the structure shown below.

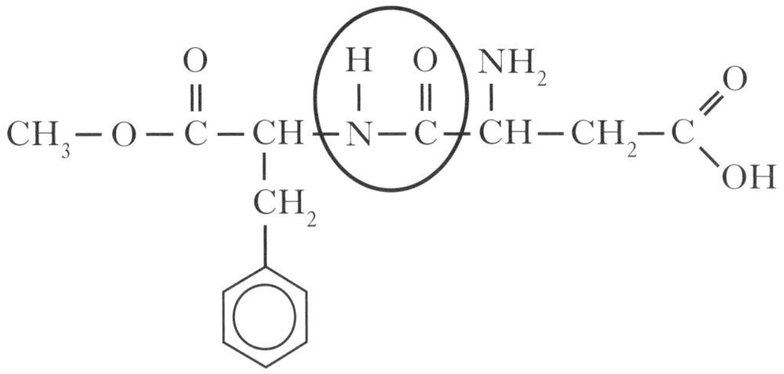

(*a*) Name the functional group circled. **1**

(*b*) In the stomach, aspartame is hydrolysed by acid to produce methanol and two amino acids, phenylalanine and aspartic acid.

Two of the products of the hydrolysis of aspartame are shown below.

CH_3-OH

methanol

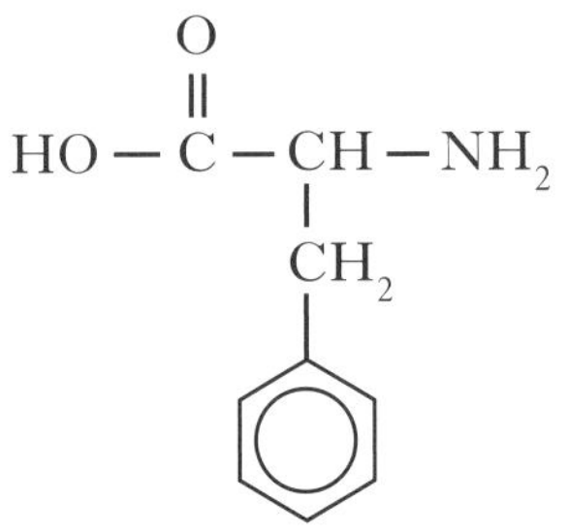

phenylalanine

Draw a structural formula for aspartic acid. **1**

DO NOT WRITE IN THIS MARGIN

Marks

8. (continued)

(*c*) The body cannot make all the amino acids it requires and is dependent on protein in the diet for the supply of certain amino acids.

What term is used to describe the amino acids the body cannot make? **1**

(*d*) To investigate this hydrolysis reaction in the lab, the apparatus shown below is set up. The extent of hydrolysis at a given temperature can be determined by measuring the quantity of methanol produced.

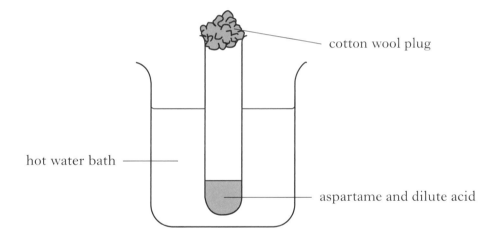

What improvement could be made to the **apparatus** to reduce the loss of methanol by evaporation? **1**

(4)

Marks

9. A fatty acid is a long chain carboxylic acid.

Examples of fatty acids are shown in the table below.

Common name	Systematic name	Structure
stearic acid	octadecanoic acid	$CH_3(CH_2)_{16}COOH$
oleic acid	octadec-9-enoic acid	$CH_3(CH_2)_7CH=CH(CH_2)_7COOH$
linoleic acid	octadec-9,12-dienoic acid	$CH_3(CH_2)_4CH=CHCH_2CH=CH(CH_2)_7COOH$
linolenic acid		$CH_3CH_2CH=CHCH_2CH=CHCH_2CH=CH(CH_2)_7COOH$

(a) Describe a chemical test, with the expected result, that could be used to distinguish between stearic and oleic acids.

1

(b) What is the systematic name for linolenic acid?

1

(c) Stearic acid can be reacted with sodium hydroxide solution to make a soap.

The structure of the soap is shown.

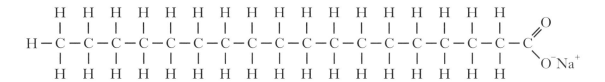

One part of the soap molecule is soluble in fat and the other part is soluble in water.

Circle the part of the soap molecule which is soluble in water.   1

(3)

Marks

10. Nitrogen and compounds containing nitrogen are widely used in industry.

(a) From which raw material is nitrogen obtained?

1

(b) In industry, methanamide, $HCONH_2$, is produced from the ester shown below.

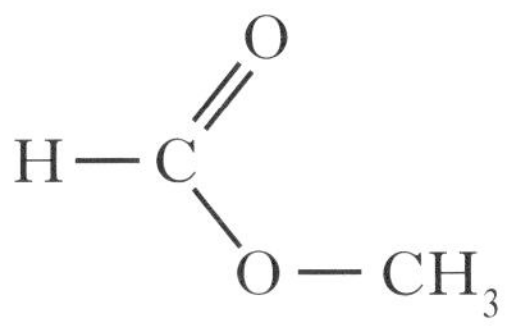

$$H-C{\overset{\displaystyle O}{\underset{\displaystyle O-CH_3}{}}}$$

Name the ester.

1

(c) In the lab, methanamide can be prepared by the reaction of methanoic acid with ammonia.

$$HCOOH + NH_3 \rightleftharpoons HCONH_2 + H_2O$$

methanoic   methanamide
acid

When 1·38 g of methanoic acid was reacted with excess ammonia, 0·945 g of methanamide was produced.

Calculate the percentage yield of methanamide.

Show your working clearly.

2

(4)

Marks

11. The element boron forms many useful compounds.

(a) Borane (BH_3) is used to synthesize alcohols from alkenes.

The reaction occurs in two stages

Stage 1 Addition Reaction

The boron atom bonds to the carbon atom of the double bond which already has the most hydrogens **directly** attached to it.

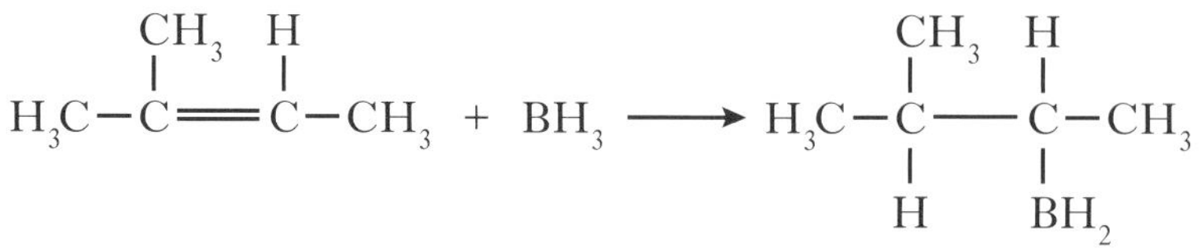

$$H_3C-C(CH_3)=C(H)-CH_3 + BH_3 \longrightarrow H_3C-C(CH_3)(H)-C(H)(BH_2)-CH_3$$

Stage 2 Oxidation Reaction

The organoborane compound is oxidised to form the alcohol.

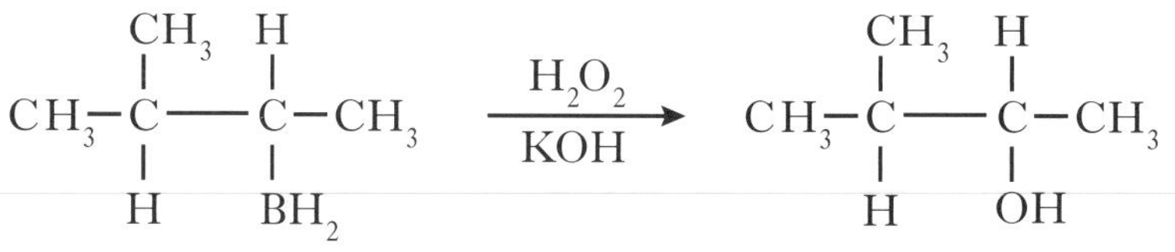

$$CH_3-C(CH_3)(H)-C(H)(BH_2)-CH_3 \xrightarrow[KOH]{H_2O_2} CH_3-C(CH_3)(H)-C(H)(OH)-CH_3$$

(i) Name the alcohol produced in Stage 2.

1

(ii) Draw a structural formula for the **alcohol** which would be formed from the alkene shown below.

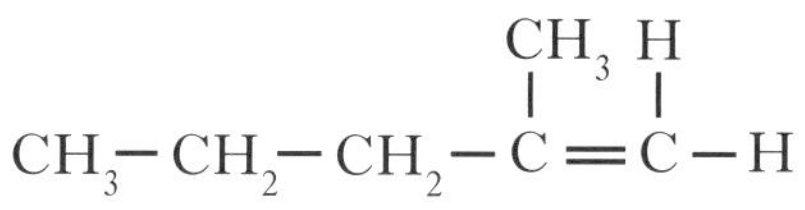

$$CH_3-CH_2-CH_2-C(CH_3)=C(H)-H$$

1

Marks

11. **(continued)**

(b) The compound diborane (B_2H_6) is used as a rocket fuel.

  (i) It can be prepared as shown.

$$BF_3 + \quad NaBH_4 \rightarrow \quad B_2H_6 + \quad NaBF_4$$

  Balance this equation.

1

  (ii) The equation for the combustion of diborane is shown below.

$$B_2H_6(g) + 3O_2(g) \rightarrow B_2O_3(s) + 3H_2O(\ell)$$

Calculate the enthalpy of combustion of diborane (B_2H_6) using the following data.

$$2B(s) + 3H_2(g) \quad \rightarrow \quad B_2H_6(g) \quad \Delta H = 36\,kJ\,mol^{-1}$$
$$H_2(g) + \tfrac{1}{2}O_2(g) \quad \rightarrow \quad H_2O(\ell) \quad \Delta H = -286\,kJ\,mol^{-1}$$
$$2B(s) + 1\tfrac{1}{2}O_2(g) \quad \rightarrow \quad B_2O_3(s) \quad \Delta H = -1274\,kJ\,mol^{-1}$$

2

(c) Diborane can be used to manufacture pentaborane (B_5H_9).

Pentaborane was also considered for use as a rocket fuel because its enthalpy of combustion is $-9037\,kJ\,mol^{-1}$.

Calculate the energy released, in kJ, when 1 kilogram of pentaborane is completely burned.

1

(6)

DO NOT
WRITE
IN THIS
MARGIN

Marks

12. In the PPA experiment **Factors Affecting Enzyme Activity**, a student investigated the effect of pH on the activity of the enzyme catalase contained in potato discs.

The following apparatus was set up and left for 3 minutes.

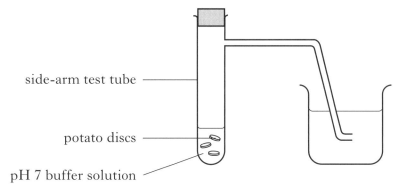

side-arm test tube

potato discs

pH 7 buffer solution

(a) Why must the buffer/potato disc mixture be left for 3 minutes before continuing the experiment?

1

(b) Which chemical must then be added to the test tube to investigate the enzyme activity at this pH?

1

(c) The activity of the enzyme was measured at several different pH values by counting the number of bubbles produced in a given time.

Why were no bubbles produced at pH 1?

1

(3)

DO NOT WRITE IN THIS MARGIN

Marks

13. Fluorine is an extremely reactive element. Its compounds are found in a range of products.

(*a*) Fluorine gas can be produced by electrolysis.

The ion-electron equation for the production of fluorine gas is:

$$2F^{-}(\ell) \rightarrow F_2(g) + 2e^{-}$$

Calculate the mass, in grams, of fluorine produced when a steady current of 5·0 A is passed through the solution for 32 minutes.

Show your working clearly. 2

[Turn over

Marks

13. **(continued)**

(b) Tetrafluoroethene, C_2F_4, is produced in industry by a series of reactions.

The final reaction in its manufacture is shown below.

$$2CHClF_2(g) \rightleftharpoons C_2F_4(g) + 2HCl(g)$$

(i) The graph shows the variation in the concentration of C_2F_4 formed at equilibrium as temperature is increased.

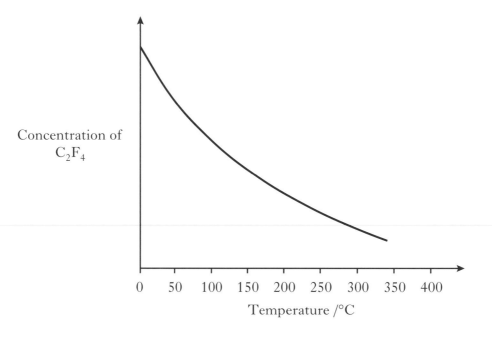

What conclusion can be drawn about the enthalpy change for the formation of tetrafluoroethene?

1

Marks

13. (b) (continued)

(ii) Sketch a graph to show how the concentration of C_2F_4 formed at equilibrium would vary with increasing pressure.

(An additional graph, if required, can be found on *Page thirty-eight*.)

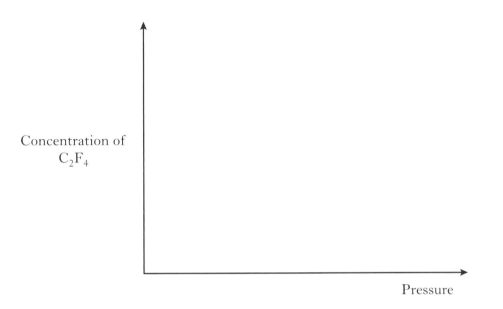

1

(c) Hydrochlorofluorocarbons are used as replacements for chlorofluorocarbons, CFCs.

What environmental problem is associated with the extensive use of CFCs?

1

(5)

[Turn over

Marks

14. Ammonium nitrate (NH_4NO_3) is widely used as a fertiliser.

   (*a*) An ammonium nitrate solution has a pH of 5.

      (i) Calculate the concentration, in mol l^{-1}, of H^+(aq) ions in the solution.

1

      (ii) **Explain clearly** why ammonium nitrate dissolves in water to produce an acidic solution.

      In your answer, you should mention the **two** equilibria involved.

2

DO NOT
WRITE
IN THIS
MARGIN

Marks

14. (continued)

(b) Ammonium nitrate must be stored and transported carefully as it can decompose according to the equation shown below:

$$2NH_4NO_3(s) \rightarrow 2N_2(g) + O_2(g) + 4H_2O(g)$$

In addition to being very exothermic, suggest another reason why the decomposition of ammonium nitrate can result in an explosion.

1

(4)

[Turn over

Marks

15. Hydrogen gas can be produced in a variety of ways.

   (*a*) Hydrogen can be produced in the lab from dilute sulphuric acid. The apparatus shown below can be used to determine the quantity of electrical charge required to form one mole of hydrogen gas.

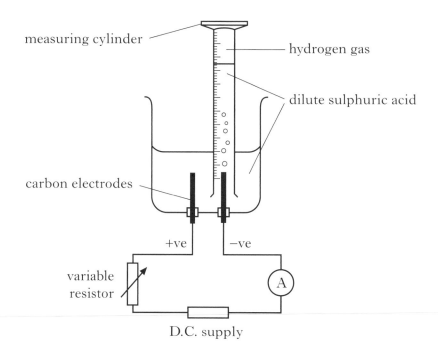

   (i) Why is a variable resistor used?

   1

   (ii) In addition to measuring the volume of hydrogen gas, which **two** other measurements must be made?

   1

Marks

15. **(continued)**

(b) The sulphur-iodine cycle is an industrial process used to manufacture hydrogen.

There are three steps in the sulphur-iodine cycle.

Step 1: $I_2 + SO_2 + 2H_2O \rightarrow 2HI + H_2SO_4$

Step 2: $2HI \rightarrow I_2 + H_2$

Step 3: $H_2SO_4 \rightarrow SO_2 + H_2O + \frac{1}{2}O_2$

   (i) Why does step 3 help to reduce the cost of manufacturing hydrogen?

1

   (ii) What is the overall equation for the sulphur-iodine cycle?

1

(4)

[Turn over

Marks

16. In alkane molecules, the chains of carbon atoms are very flexible. The molecules can twist at any carbon-to-carbon bond.

Butane is a typical alkane.

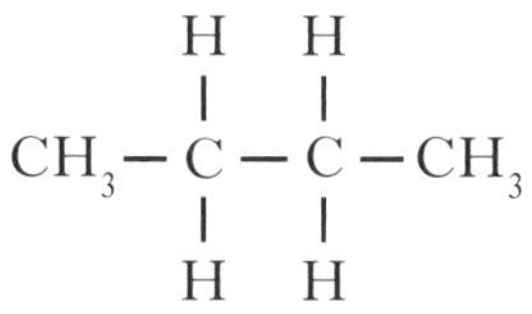

$$CH_3 - \overset{\displaystyle H}{\underset{\displaystyle H}{C}} - \overset{\displaystyle H}{\underset{\displaystyle H}{C}} - CH_3$$

The diagram below shows butane twisting about the central carbon-to-carbon bond.

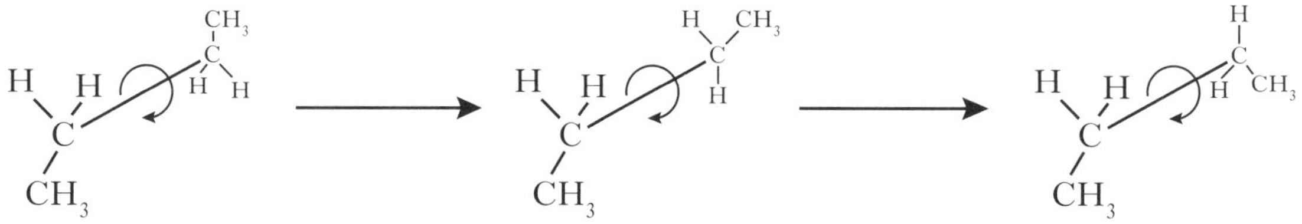

Newman projections are special diagrams used to show the relative position of the different atoms in a molecule.

In the diagram below, a Newman projection has been drawn showing the relative position of the atoms in the butane molecule below.

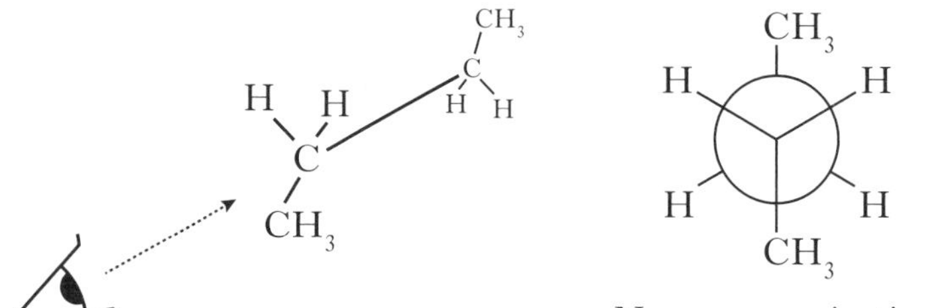

Newman projection

(*a*) Complete the Newman projection diagram for the butane molecule in the position shown below.

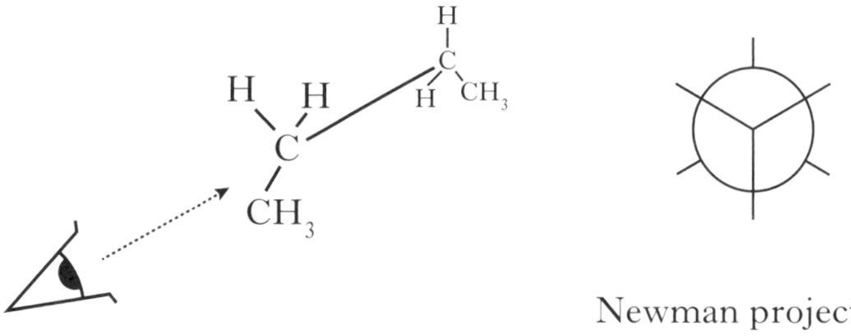

Newman projection   1

Marks

16. **(continued)**

(*b*) Name the alkane molecule represented by the Newman projection shown below.

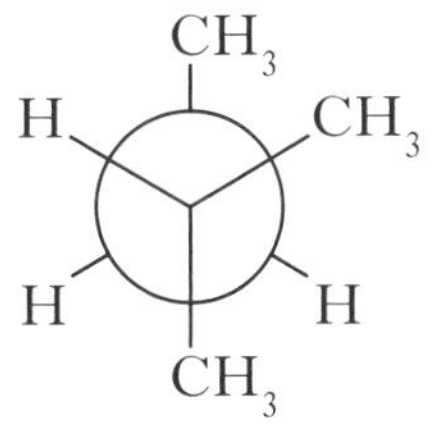

1

(2)

[*END OF QUESTION PAPER*]

ADDITIONAL GRAPH FOR USE IN QUESTION 13(b)(ii)

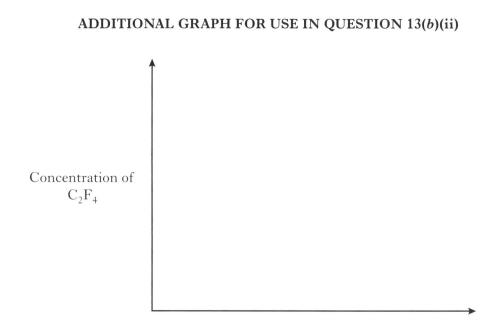

ADDITIONAL SPACE FOR ANSWERS

[BLANK PAGE]

HIGHER

2013

[BLANK PAGE]

FOR OFFICIAL USE

Total
Section B

X012/12/02

NATIONAL	FRIDAY, 31 MAY	CHEMISTRY
QUALIFICATIONS	1.00 PM – 3.30 PM	HIGHER
2013		

Fill in these boxes and read what is printed below.

Full name of centre

Town

Forename(s)

Surname

Date of birth

Day	Month	Year	Scottish candidate number	Number of seat

Reference may be made to the Chemistry Higher and Advanced Higher Data Booklet.

SECTION A—Questions 1–40 (40 marks)

Instructions for completion of **Section A** are given on page two.

For this section of the examination you must use an **HB pencil**.

SECTION B (60 marks)

1   All questions should be attempted.

2   The questions may be answered in any order but all answers are to be written in the spaces provided in this answer book, **and must be written clearly and legibly in ink**.

3   Rough work, if any should be necessary, should be written in this book and then scored through when the fair copy has been written. If further space is required, a supplementary sheet for rough work may be obtained from the Invigilator.

4   Additional space for answers will be found at the end of the book. If further space is required, supplementary sheets may be obtained from the Invigilator and should be inserted inside the **front** cover of this book.

5   The size of the space provided for an answer should not be taken as an indication of how much to write. It is not necessary to use all the space.

6   Before leaving the examination room you must give this book to the Invigilator. If you do not, you may lose all the marks for this paper.

SECTION A

Read carefully

1 Check that the answer sheet provided is for **Chemistry Higher (Section A)**.

2 For this section of the examination you must use an **HB pencil** and, where necessary, an eraser.

3 Check that the answer sheet you have been given has **your name**, **date of birth**, **SCN** (Scottish Candidate Number) and **Centre Name** printed on it.

   Do not change any of these details.

4 If any of this information is wrong, tell the Invigilator immediately.

5 If this information is correct, **print** your name and seat number in the boxes provided.

6 The answer to each question is **either** A, B, C or D. Decide what your answer is, then, using your pencil, put a horizontal line in the space provided (see sample question below).

7 There is only **one correct answer** to each question.

8 Any rough working should be done on the question paper or the rough working sheet, **not** on your answer sheet.

9 At the end of the examination, put the **answer sheet for Section A inside the front cover of your answer book**.

Sample Question

To show that the ink in a ball-pen consists of a mixture of dyes, the method of separation would be

     A chromatography

     B fractional distillation

     C fractional crystallisation

     D filtration.

The correct answer is **A**—chromatography. The answer **A** has been clearly marked in **pencil** with a horizontal line (see below).

Changing an answer

If you decide to change your answer, carefully erase your first answer and using your pencil, fill in the answer you want. The answer below has been changed to **D**.

1. Which of the following chlorides conducts electricity when molten?

   A   Calcium chloride

   B   Nitrogen chloride

   C   Phosphorus chloride

   D   Silicon chloride

2. Particles with the same electron arrangement are said to be isoelectronic.

   Which of the following compounds contains ions which are isoelectronic?

   A   CaO

   B $CaBr_2$

   C Na_2O

   D   LiF

3. Which of the following would be expected to react?

   A   Iron and zinc sulphate solution

   B   Tin and silver nitrate solution

   C   Copper and dilute sulphuric acid

   D   Lead and magnesium chloride solution

4. An excess of silver nitrate solution is added to a solution of sodium chloride, and a white precipitate is formed. The precipitate is then filtered off.

   Which solution would **not** give a precipitate when added to the filtrate?

   A   Barium chloride

   B   Potassium nitrate

   C   Calcium iodide

   D   Sodium bromide

   (You may wish to refer to the Data Booklet.)

5. Excess marble chips (calcium carbonate) were added to $25\,cm^3$ of hydrochloric acid, concentration $2\,mol\,l^{-1}$.

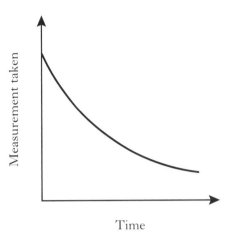

   Which of the following measurements, taken at regular intervals and plotted against time, would give the graph shown above?

   A   Temperature

   B   Volume of gas produced

   C   pH of solution

   D   Mass of the beaker and contents

6. In which of the following will **both** changes result in an increase in the rate of a chemical reaction?

   A   A decrease in activation energy and an increase in the frequency of collisions

   B   An increase in activation energy and a decrease in particle size

   C   An increase in temperature and an increase in the particle size

   D   An increase in concentration and a decrease in the surface area of the reactant particles

[Turn over

7. When $50 \, cm^3$ of $2 \, mol \, l^{-1}$ hydrochloric acid was added to $50 \, cm^3$ of $1 \, mol \, l^{-1}$ sodium hydroxide, the temperature increased by $6 \, °C$.

   Which of the following changes to one of the reactants would bring about a greater increase in temperature, assuming the other reactant was not changed?

   A   Use $500 \, cm^3$ of $2 \, mol \, l^{-1}$ hydrochloric acid.

   B   Use $50 \, cm^3$ of $4 \, mol \, l^{-1}$ hydrochloric acid.

   C   Use $50 \, cm^3$ of $1 \, mol \, l^{-1}$ potassium hydroxide.

   D   Use $50 \, cm^3$ of $2 \, mol \, l^{-1}$ sodium hydroxide.

8. Which of the following statements is **not** correct?

   A   The surface activity of a catalyst can be reduced by poisoning.

   B   Impurities in the reactants can result in a catalyst having to be regenerated.

   C   Homogenous catalysts are found in the catalytic converters fitted to cars.

   D   Heterogenous catalysts work by adsorbing reactant molecules.

9. Which of the following diagrams represents an exothermic reaction which is most likely to take place at room temperature?

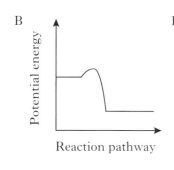

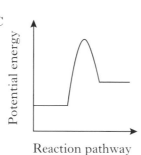

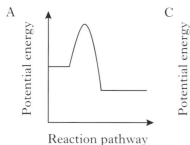

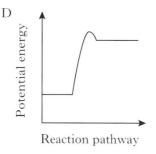

10. The enthalpy of combustion of methanol is $-727 \, kJ \, mol^{-1}$.

    What mass of methanol has to be burned to produce $72.7 \, kJ$?

    A $3.2 \, g$

    B $32 \, g$

    C $72.7 \, g$

    D $727 \, g$

11. Which of the following elements has the greatest attraction for bonding electrons?

    A   Lithium

    B   Chlorine

    C   Sodium

    D   Bromine

12. Which of the following statements is true?

    A   The potassium ion is larger than the potassium atom.

    B   The chloride ion is smaller than the chlorine atom.

    C   The sodium atom is larger than the sodium ion.

    D   The oxygen atom is larger than the oxide ion.

13. The diagram shows the melting points of successive elements across a period in the Periodic Table.

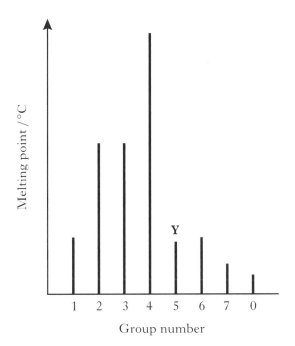

Which of the following is a correct reason for the low melting point of element **Y**?

A   It has weak ionic bonds.

B   It has weak covalent bonds.

C   It has weakly-held outer electrons.

D   It has weak forces between molecules.

14. Which type of bonding is **never** found in elements?

A   Metallic

B   van der Waals'

C   Polar covalent

D   Non-polar covalent

15. Which of the following elements exists as discrete molecules?

A   Boron

B   Carbon (diamond)

C   Silicon

D   Sulphur

16. The structures for molecules of four liquids are shown below.

Which liquid will be the most viscous?

A

$$H-\overset{\overset{\displaystyle H}{|}}{\underset{\underset{\displaystyle H}{|}}{C}}-\overset{\overset{\displaystyle H}{|}}{\underset{\underset{\displaystyle H}{|}}{C}}-\overset{\overset{\displaystyle H}{|}}{\underset{\underset{\displaystyle H}{|}}{C}}-\overset{\overset{\displaystyle H}{|}}{\underset{\underset{\displaystyle H}{|}}{C}}-OH$$

B

$$H-\overset{\overset{\displaystyle H}{|}}{\underset{\underset{\displaystyle H}{|}}{C}}-\overset{\overset{\displaystyle H}{|}}{\underset{\underset{\displaystyle H}{|}}{C}}-\overset{\overset{\displaystyle H}{|}}{\underset{\underset{\displaystyle OH}{|}}{C}}-\overset{\overset{\displaystyle H}{|}}{\underset{\underset{\displaystyle H}{|}}{C}}-H$$

C

$$H-\overset{\overset{\displaystyle H}{|}}{\underset{\underset{\displaystyle H}{|}}{C}}-\overset{\overset{\displaystyle \overset{\displaystyle H-C-H}{|}}{|}}{\underset{\underset{\displaystyle OH}{|}}{C}}-\overset{\overset{\displaystyle H}{|}}{\underset{\underset{\displaystyle H}{|}}{C}}-H$$

D

$$H-\overset{\overset{\displaystyle H}{|}}{\underset{\underset{\displaystyle OH}{|}}{C}}-\overset{\overset{\displaystyle H}{|}}{\underset{\underset{\displaystyle H}{|}}{C}}-\overset{\overset{\displaystyle H}{|}}{\underset{\underset{\displaystyle H}{|}}{C}}-\overset{\overset{\displaystyle H}{|}}{\underset{\underset{\displaystyle OH}{|}}{C}}-H$$

17. A mixture of magnesium bromide and magnesium sulphate is known to contain 3 mol of magnesium and 4 mol of bromide ions.

How many moles of sulphate ions are present?

A   1

B   2

C   3

D   4

[Turn over

18.

$$2C_2H_2(g) + 5O_2(g) \longrightarrow 4CO_2(g) + 2H_2O(\ell)$$
ethyne

What volume of gas would be produced by the complete combustion of $100\,cm^3$ of ethyne gas?

All volumes were measured at atmospheric pressure and room temperature.

A $200\,cm^3$

B $300\,cm^3$

C $400\,cm^3$

D $800\,cm^3$

19. Which equation could represent an industrial cracking process?

A $CH_3(CH_2)_6CH_2OH$
$\downarrow$
$CH_3(CH_2)_5 CH = CH_2 + H_2O$

B $CH_3(CH_2)_6 CH_3$
$\downarrow$
$CH_3C(CH_3)_2CH_2CH(CH_3)_2$

C $CH_3(CH_2)_6CH_3$
$\downarrow$
$CH_3(CH_2)_4CH_3 + CH_2 = CH_2$

D $4CH_2 = CH_2 \rightarrow \quad -(CH_2CH_2)_4-$

20. Which of the following hydrocarbons is an isomer of 2-methylpent-2-ene?

A
$$\begin{array}{c} CH_3 \\ | \\ C = CH \\ | \quad\quad | \\ CH_3 \quad CH_2CH_3 \end{array}$$

B
$$\begin{array}{c} \quad\quad\quad CH_3 \\ \quad\quad\quad | \\ CH = C \\ | \quad\quad | \\ CH_3 \quad CH_2CH_3 \end{array}$$

C
$$\begin{array}{c} CH_3 \\ | \\ C = CH - CH_2 - CH_3 \\ | \\ CH_3 \end{array}$$

D
$$\begin{array}{c} CH_3 \quad\quad\quad CH_3 \\ | \quad\quad\quad\quad | \\ CH_2 - CH = C \\ \quad\quad\quad\quad | \\ \quad\quad\quad\quad CH_3 \end{array}$$

21. A compound with molecular formula $C_6H_{12}O_2$, could be

A hexanal

B hexan-2-ol

C hexan-2-one

D hexanoic acid.

22. An ester has the following structural formula
$CH_3CH_2CH_2COOCH_2CH_3$

The name of this ester is

A propyl propanoate

B ethyl butanoate

C butyl ethanoate

D ethyl propanoate.

23. Cyclohexylethene and phenylethene are important industrial feedstocks.

CH=CH$_2$
|
CH
H$_2$C   CH$_2$
H$_2$C   CH$_2$
CH$_2$

CH=CH$_2$

cyclohexylethene   phenylethene

Phenylethene

A   cannot undergo addition polymerisation but cyclohexylethene can

B   undergoes addition reactions much more quickly than cyclohexylethene

C   contains 5 fewer hydrogen atoms per molecule than cyclohexylethene

D   decolourises the same number of moles of bromine as cyclohexylethene.

24. The product formed when propan–1–ol is dehydrated is

A   propane

B   propene

C   propanal

D   propanoic acid.

25. Which **two** isomers would each produce an acid when warmed with acidified potassium dichromate solution?

1    $CH_3-CH_2-CH_2-CH_2-OH$

2    $CH_3-CH-CH_2-CH_3$
               |
              OH

3
                 CH$_3$
                 |
         CH_3-C-OH
                 |
                 CH$_3$

4    $CH_3-CH-CH_3$
              |
             CH_2-OH

A   1 and 2

B   2 and 3

C   1 and 4

D   3 and 4

26. Esters are formed by the reaction between which **two** functional groups?

A   A hydroxyl group and a carboxyl group

B   A hydroxyl group and a carbonyl group

C   A hydroxide group and a carboxyl group

D   A hydroxide group and a carbonyl group

27. Polyamides and polyesters are always made from monomers

A   which are unsaturated

B   with one functional group per molecule

C   containing a benzene ring structure

D   with two functional groups per molecule.

28. The process of cross-linking occurs in the

A   conversion of vegetable oils to margarine

B   curing of polyester resins

C   production of aromatic compounds from naphtha

D   manufacture of thermoplastics.

29. Oils are generally

    A   liquid at room temperature and contain a high proportion of unsaturated molecules

    B   liquid at room temperature and contain a high proportion of saturated molecules

    C   solid at room temperature and contain a high proportion of unsaturated molecules

    D   solid at room temperature and contain a high proportion of saturated molecules.

30. A tripeptide **X** has the structure

$$\underset{H_2N-CH-CONH-CH_2-CONH-CH-COOH}{\overset{CH_3 \qquad\qquad\qquad\qquad CH(CH_3)_2}{|\qquad\qquad\qquad\qquad\qquad\qquad\quad|}}$$

    Partial hydrolysis of **X** yields a mixture of dipeptides.

    Which of the following dipeptides could be produced on hydrolysing **X**?

    A
$$\underset{H_2N-CH_2-CONH-CH-COOH}{\overset{CH(CH_3)_2}{\qquad\qquad\qquad\qquad|}}$$

    B
$$\underset{H_2N-CH_2-CONH-CH-COOH}{\overset{CH_3}{\qquad\qquad\qquad\qquad|}}$$

    C
$$\underset{H_2N-CH-CONH-CH-COOH}{\overset{CH_3 \qquad\quad CH(CH_3)_2}{|\qquad\qquad\qquad\quad|}}$$

    D
$$\underset{H_2N-CH-CONH-CH_2-COOH}{\overset{CH(CH_3)_2}{|}}$$

31. The costs involved in the industrial production of a chemical are made up of fixed costs and variable costs.

    Which of the following is most likely to be classified as a variable cost?

    A   The cost of labour

    B   The cost of land rental

    C   The cost of raw materials

    D   The cost of plant construction

32.

$$C(s) + H_2(g) + O_2(g) \rightarrow HCOOH(\ell) \qquad \Delta H = a$$

$$HCOOH(\ell) + {}^1/_2O_2(g) \rightarrow CO_2(g) + H_2O(\ell) \qquad \Delta H = b$$

$$C(s) + O_2(g) \rightarrow CO_2(g) \qquad \Delta H = c$$

$$H_2(g) + {}^1/_2O_2(g) \rightarrow H_2O(\ell) \qquad \Delta H = d$$

    What is the relationship between a, b, c and d?

    A $a = c + d - b$

    B $a = b - c - d$

    C $a = -b - c - d$

    D $a = c + b + d$

33. $2SO_2(g) + O_2(g) \rightleftharpoons 2SO_3(g)$

    The equation represents a mixture at equilibrium.

    Which line in the table is true for the mixture after a further 2 hours of reaction?

	Rate of forward reaction	Rate of back reaction
A	unchanged	unchanged
B	increases	increases
C	decreases	decreases
D	unchanged	decreases

34. In which of the following would an increase in pressure result in the equilibrium position being moved to the left?

    A $N_2(g) + 3H_2(g) \rightleftharpoons 2NH_3(g)$

    B $CO(g) + H_2O(g) \rightleftharpoons CO_2(g) + H_2(g)$

    C $CH_4(g) + H_2O(g) \rightleftharpoons CO(g) + 3H_2(g)$

    D $Fe_2O_3(s) + 3CO(g) \rightleftharpoons 2Fe(s) + 3CO_2(g)$

35. Which of the following is the same for equal volumes of $0 \cdot 1 \, mol \, l^{-1}$ solutions of sodium hydroxide and ammonia?

    A   The pH of solution

    B   The mass of solute present

    C   The conductivity of solution

    D   The number of moles of hydrochloric acid needed for neutralisation

36. The concentration of $OH^-(aq)$ ions in a solution is $0 \cdot 1\ mol\ l^{-1}$.

    What is the pH of the solution?

    A   1

    B   8

    C   13

    D   14

37. Which of the following $1\ mol\ l^{-1}$ solutions has a pH greater than 1 but less than 7?

    A $HCl(aq)$

    B $NaCl(aq)$

    C $Na_2CO_3(aq)$

    D $CH_3COOH(aq)$

38. During electrolysis, which of the following would be formed by the passage of $96\,500$ C?

    A $0 \cdot 5\ mol\ Ag$ from $AgNO_3(aq)$

    B $1 \cdot 0\ mol\ Ag$ from $AgNO_3(aq)$

    C $1 \cdot 0\ mol\ Cu$ from $CuSO_4(aq)$

    D $2 \cdot 0\ mol\ Cu$ from $CuSO_4(aq)$

39. Identify the graph which shows how rate varies with temperature for the radioactive decay of uranium-235.

    A
    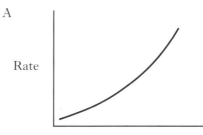

    B
    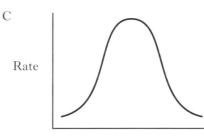

    C
    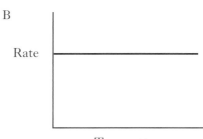

    D
    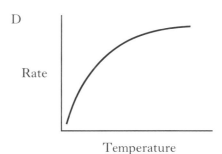

40. Which of the following equations represents a nuclear fission process?

    A $^{31}_{15}P + {}^{1}_{0}n \rightarrow {}^{32}_{15}P$

    B $^{2}_{1}H + {}^{3}_{1}H \rightarrow {}^{4}_{2}He + {}^{1}_{0}n$

    C $^{235}_{92}U + {}^{1}_{0}n \rightarrow {}^{90}_{38}Sr + {}^{144}_{54}Xe + 2{}^{1}_{0}n$

    D $^{27}_{13}Al + {}^{4}_{2}He \rightarrow {}^{30}_{15}P + {}^{1}_{0}n$

Candidates are reminded that the answer sheet MUST be returned INSIDE the front cover of this answer book.

[BLANK PAGE]

Marks

SECTION B

All answers must be written clearly and legibly in ink.

1. An example of a reaction used to produce hydrocarbons for use in unleaded petrol is shown.

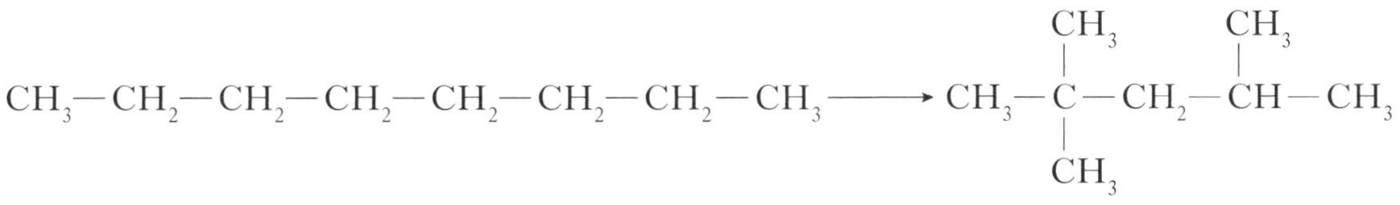

$$CH_3-CH_2-CH_2-CH_2-CH_2-CH_2-CH_2-CH_3 \longrightarrow CH_3-\overset{\overset{\displaystyle CH_3}{|}}{\underset{\underset{\displaystyle CH_3}{|}}{C}}-CH_2-\overset{\overset{\displaystyle CH_3}{|}}{CH}-CH_3$$

octane   compound A

(a) Name the type of reaction shown above.

1

(b) Give the systematic name for compound **A**.

1

(c) What structural feature of compound **A** makes it suitable for use in unleaded petrol?

1

(d) Methanol can be used as an alternative fuel to petrol.

State one disadvantage of using methanol as a fuel.

1

(4)

[Turn over

Marks

2. A student carries out a Prescribed Practical Activity (PPA) to determine the effect of temperature on the rate of the reaction between oxalic acid and acidified potassium permanganate solution.

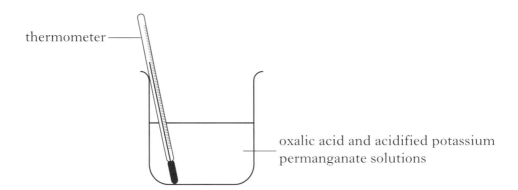

thermometer

oxalic acid and acidified potassium permanganate solutions

(a) What colour change indicates that the reaction is complete?

1

(b) The student's results are shown on the graph below.

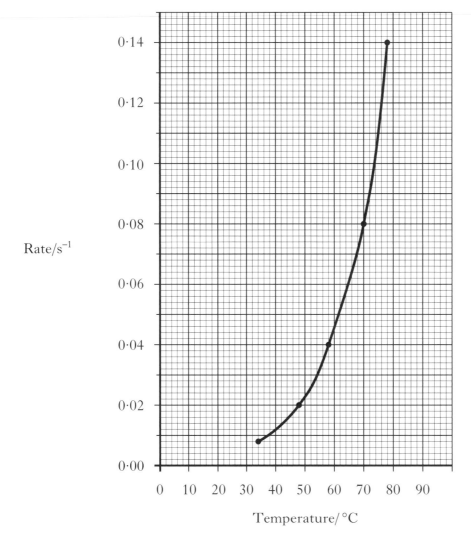

Rate/s⁻¹

Temperature/°C

DO NOT WRITE IN THIS MARGIN

Marks

2. (*b*) (continued)

(i) The reaction time recorded in one experiment was 25 s.

Use the graph to determine the temperature, in °C, of this reaction. **1**

(ii) Why is it difficult to obtain an accurate reaction time when the reaction is carried out below 30 °C? **1**

(*c*) A small increase in temperature can cause a large increase in reaction rate.

As temperature is increased, collisions occur more frequently.

What other reason is there for the large increase in reaction rate observed when the temperature is increased? **1**

(4)

[Turn over

DO NOT WRITE IN THIS MARGIN

Marks

3. Ammonium perchlorate, NH_4ClO_4, is used in solid fuel rocket boosters.

(*a*) In the rocket boosters, ammonium perchlorate reacts with aluminium as shown.

$$Al + NH_4ClO_4 \rightarrow Al_2O_3 + AlCl_3 + NO + H_2O$$

Balance this equation. **1**

(*b*) Calculate the mass of aluminium oxide, in g, that would contain $3{\cdot}01 \times 10^{21}$ aluminium ions.

(The mass of one mole of Al_2O_3 = 102·0 g).

Show your working clearly. **2**

(3)

DO NOT WRITE IN THIS MARGIN

Marks

4. Attempts have been made to make foods healthier by using alternatives to traditional cooking ingredients.

(*a*) An alternative to common salt contains potassium ions and chloride ions.

(i) Write an ion-electron equation for the first ionisation energy of potassium. **1**

(ii) **Explain clearly** why the first ionisation energy of potassium is smaller than that of chlorine. **2**

(*b*) A calorie-free replacement for fat can be made by reacting fatty acids with the hydroxyl groups on a molecule of sucrose. A structural formula for sucrose is shown.

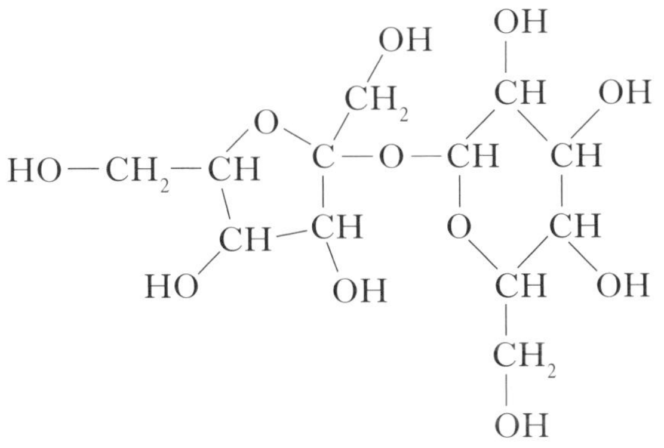

How many fatty acid molecules can react with one molecule of sucrose? **1**

(4)

5. Cyanogen gas, $C_2N_2(g)$, is a compound of carbon and nitrogen.

    (*a*) Draw a full structural formula for a cyanogen molecule.

1

    (*b*) In a combustion chamber, cyanogen gas burns to form a mixture of carbon dioxide and nitrogen.

$$C_2N_2(g) \ + \ 2O_2(g) \ \rightarrow \ 2CO_2(g) \ + \ N_2(g)$$

    Carbon dioxide can be removed by passing the gas mixture through sodium hydroxide solution.

    Complete the diagram to show how carbon dioxide can be removed from the products and the volume of nitrogen gas measured.

    (An additional diagram, if required, can be found on *Page thirty-four*.)

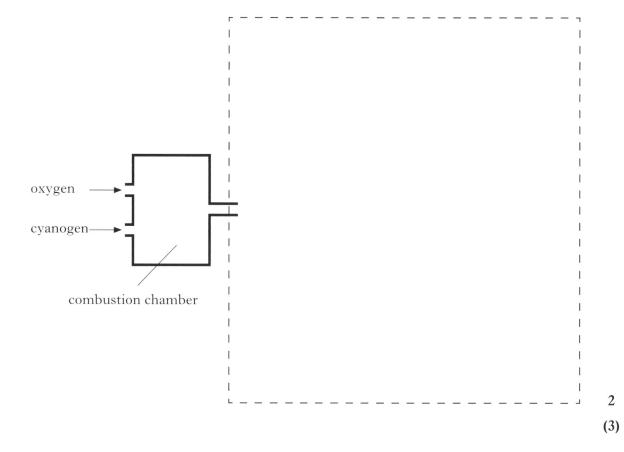

2

(3)

DO NOT WRITE IN THIS MARGIN

Marks

6. The structures below show molecules that contain chlorine atoms.

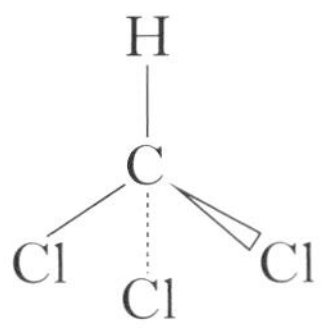

trichloromethane

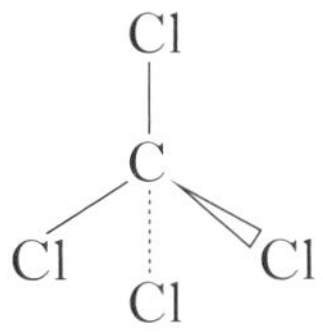

tetrachloromethane

(*a*) **Explain clearly** why trichloromethane is more soluble in water than tetrachloromethane.

Your answer should include the names of the intermolecular forces involved. **2**

(*b*) Tetrachloromethane damages the Earth's ozone layer.

Why is the ozone layer important? **1**

(3)

Marks

7. Many of the flavour and aroma molecules found in chocolate are aldehydes and ketones.

Two examples are shown below.

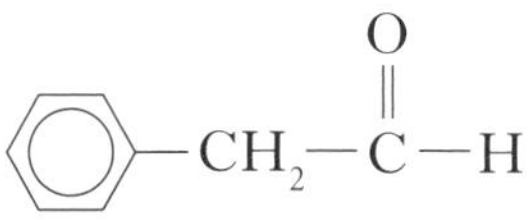

phenylethanal 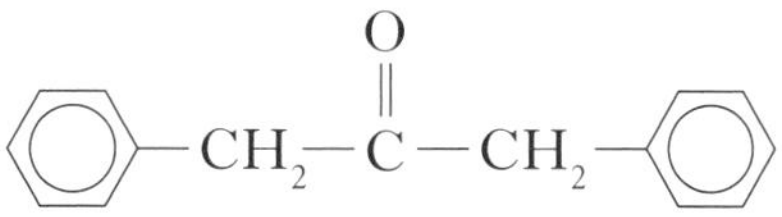   1,3-diphenylpropan-2-one

(a) Name a chemical that could be used to distinguish between these two compounds.

1

(b)   (i) How many hydrogen atoms are present in a molecule of phenylethanal?

1

(ii) Phenylethanal can be converted to phenylethanoic acid.

Name the type of chemical reaction taking place.

1

(3)

DO NOT WRITE IN THIS MARGIN

Marks

8. Poly(vinyl carbazole) is a useful polymer.

The structure of a section of the polymer is shown below.

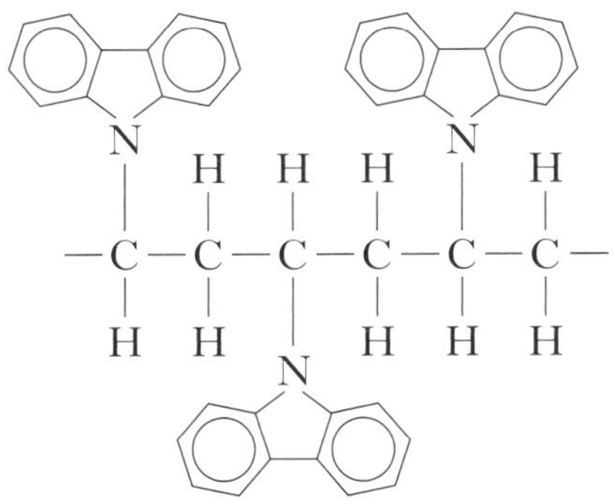

(*a*) Draw a structural formula for the vinyl carbazole monomer used to make poly(vinyl carbazole). **1**

(*b*) Give a use for poly(vinyl carbazole). **1**

(2)

DO NOT WRITE IN THIS MARGIN

Marks

9. Alcohols are widely used in antifreeze and de-icers.

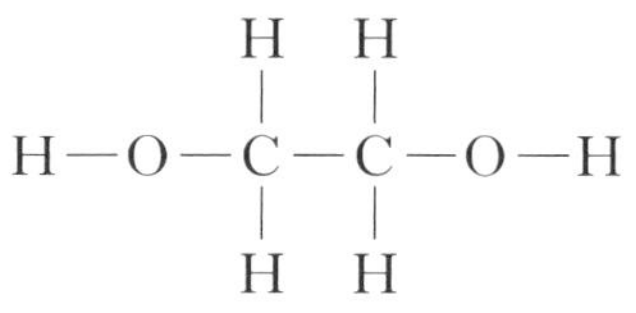

ethane-1,2-diol

molecular mass 62

boiling point 197 °C

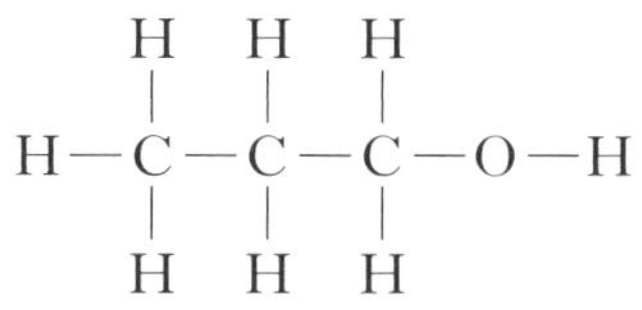

propan-1-ol

molecular mass 60

boiling point 98 °C

(*a*) Why is the boiling point of ethane-1,2-diol much higher than the boiling point of propan-1-ol? **1**

(*b*) Ethane-1,2-diol can be produced industrially from ethene in a two stage process:

Stage one

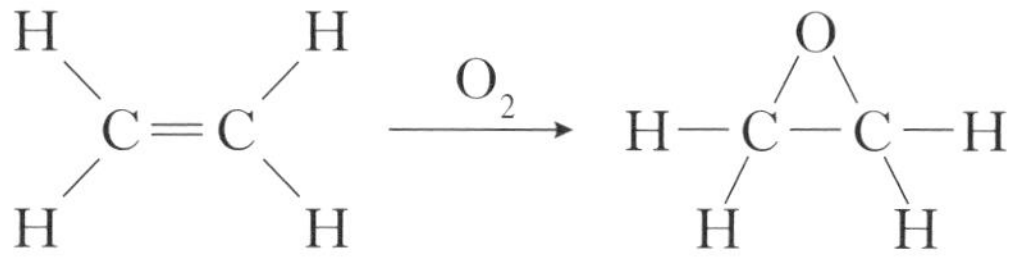

Stage two

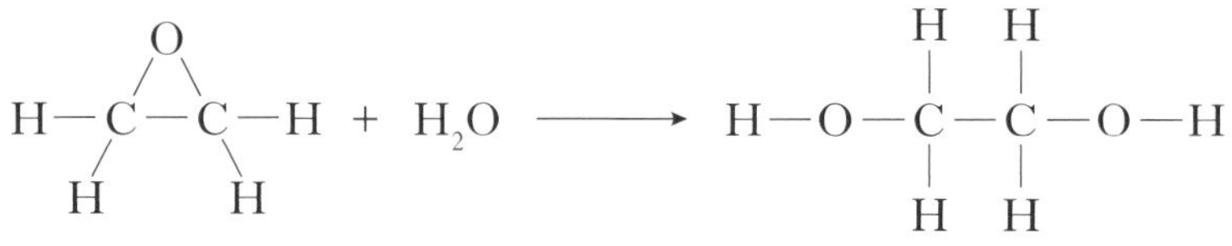

Name the alkene required to produce 2-methylbutane-2,3-diol. **1**

Marks

9. **(continued)**

(c) Ethane-1,2-diol can react with benzene-1,4-dicarboxylic acid to produce a polymer.

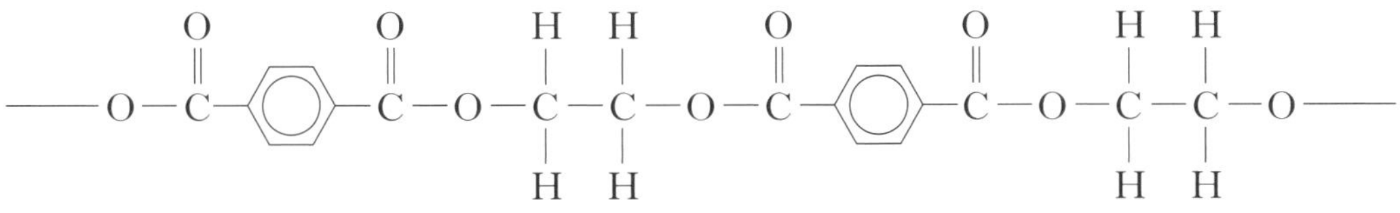

Name the acid which, when added to ethane-1,2-diol, would produce the polymer shown below.

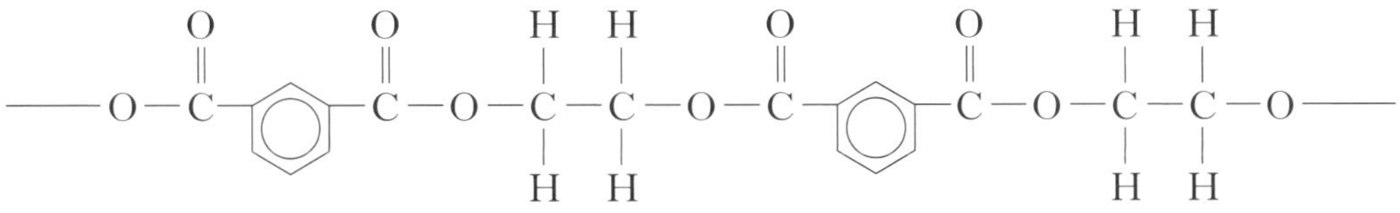

1

(3)

[Turn over

DO NOT WRITE IN THIS MARGIN

Marks

10. Dental anaesthetics are substances used to reduce discomfort during treatment.

(*a*) Procaine is a dental anaesthetic.

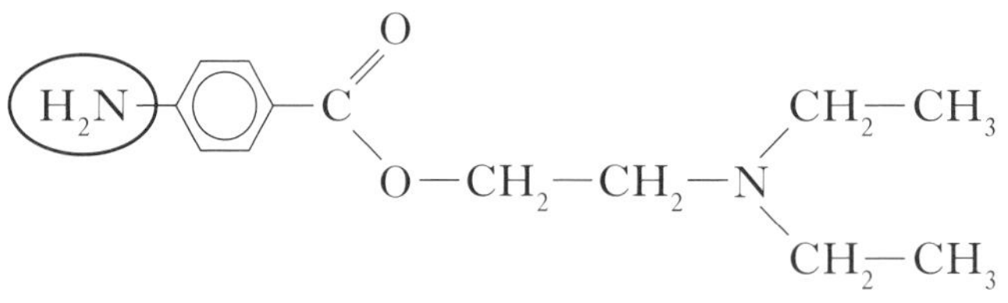

(i) Name the functional group circled above. **1**

(ii) Procaine causes numbness when applied to the gums. This effect wears off as the procaine is hydrolysed.

One of the products of the hydrolysis of procaine is shown below.

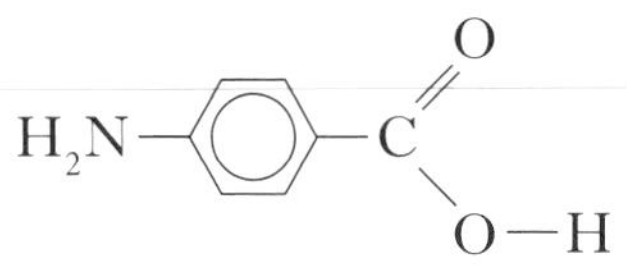

Draw a structural formula for the other compound produced when procaine is hydrolysed. **1**

Marks

10. (continued)

(b)   The table below shows the duration of numbness for some anaesthetics.

Name of anaesthetic	Structure	Duration of numbness/ minutes
procaine	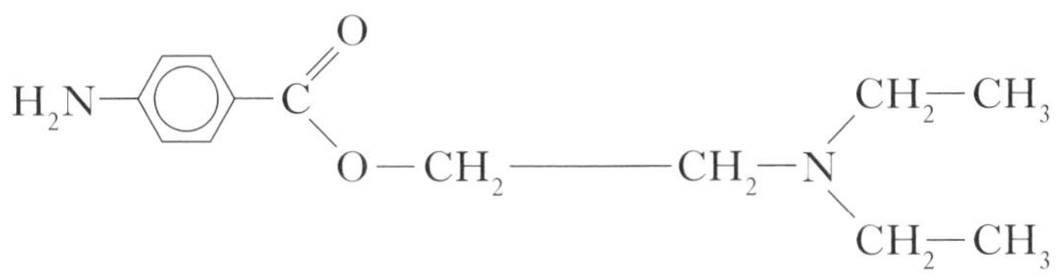	7
lidocaine	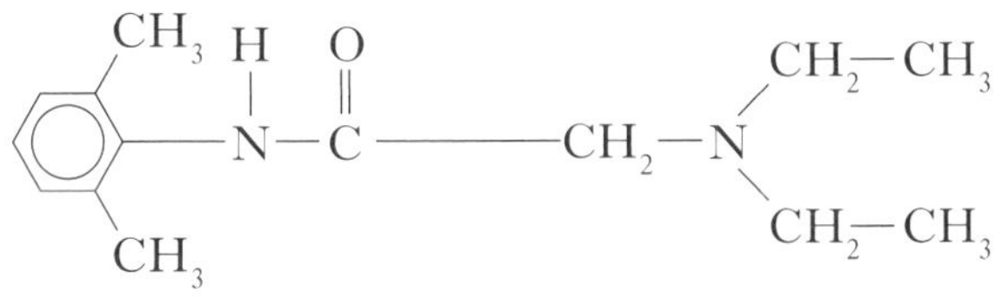	96
mepivacaine	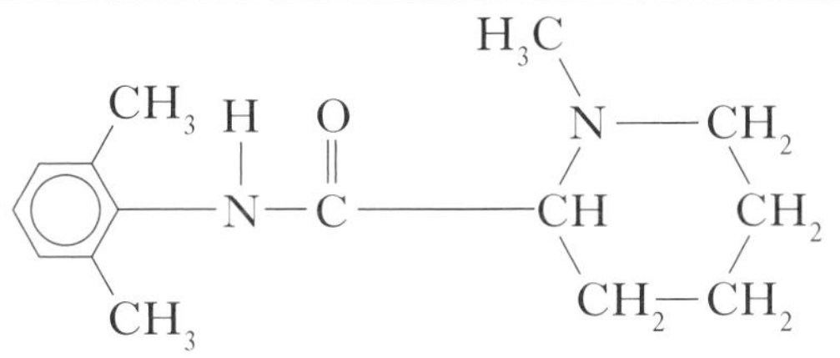	114
anaesthetic **X**	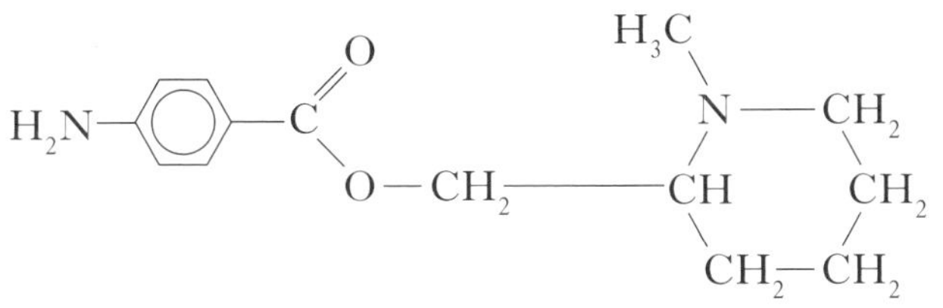	

Estimate the duration of numbness, in minutes, for anaesthetic **X**.

1

(3)

[Turn over

Marks

11. Aspirin, a common painkiller, can be made by the reaction of salicylic acid with ethanoic anhydride.

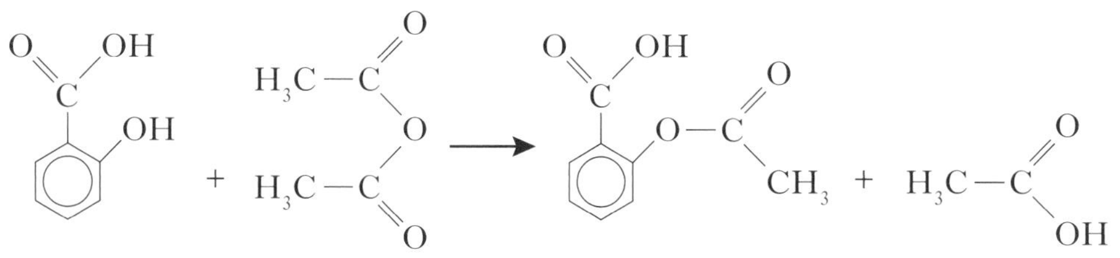

| $C_7H_6O_3$ | $C_4H_6O_3$ | $C_9H_8O_4$ | $C_2H_4O_2$ |
| salicylic acid | ethanoic anhydride | aspirin | ethanoic acid |

(a) Name the type of reaction that takes place in the formation of aspirin from salicylic acid and ethanoic anhydride.

1

(b) In a laboratory preparation of aspirin, 5·02 g of salicylic acid produced 2·62 g of aspirin.

Calculate the percentage yield of aspirin.

Show your working clearly.

2

(c) The sodium salt of aspirin is more soluble in water than aspirin itself.

Why does a solution of the sodium salt of aspirin have an alkaline pH?

1

(4)

DO NOT WRITE IN THIS MARGIN

Marks

12. The age of a rock found in Canada was determined by measuring the amounts of argon-40 and potassium-40 present in a sample.

(*a*) Each potassium-40 atom can emit a single positron particle to form an argon-40 atom.

Complete the table below to show the mass number and atomic number for a positron.

(An additional table, if required, can be found on *Page thirty-four*.)

Mass number	
Atomic number	

1

(*b*) 75% of the potassium-40 atoms originally present in the rock sample were found to have undergone radioactive decay.

The half-life of potassium-40 is $1{\cdot}26 \times 10^9$ years.

Calculate the age of the rock, in years.

1

(2)

[Turn over

Marks

13. Hydrogen sulphide is a toxic gas with the smell of rotten eggs.

   (a) Hydrogen sulphide dissolves readily in water:

   $$H_2S(aq) \rightleftharpoons HS^-(aq) + H^+(aq)$$

   (i) Why can hydrogen sulphide be described as a weak acid?

   1

   (ii) What effect would the addition of ammonia solution have on the position of equilibrium in the above reaction?

   1

   (b) Hydrogen sulphide gas can be prepared by the reaction of iron(II) sulphide with excess dilute hydrochloric acid:

   $$FeS(s) + 2HCl(aq) \rightarrow FeCl_2(aq) + H_2S(g)$$

   Calculate the mass of iron(II) sulphide required to produce 79 cm³ of hydrogen sulphide gas.

   (Take the molar volume of hydrogen sulphide to be 24 litres mol⁻¹.)

   Show your working clearly.

   2

   (4)

Marks

14. Mobile phones are being developed that can be powered by methanol.

Methanol can be made by a two-stage process.

(*a*) In the first stage, methane is reacted with steam to produce a mixture of carbon monoxide and hydrogen.

$$CH_4(g) + H_2O(g) \rightleftharpoons CO(g) + 3H_2(g)$$

  (i) Give the name for the mixture of carbon monoxide and hydrogen which is produced.

1

  (ii) Use the data below to calculate the enthalpy change, in $kJ\,mol^{-1}$, for the forward reaction.

$$CO(g) + \tfrac{1}{2}O_2(g) \rightarrow CO_2(g) \qquad \Delta H = -283\,kJ\,mol^{-1}$$
$$H_2(g) + \tfrac{1}{2}O_2(g) \rightarrow H_2O(g) \qquad \Delta H = -242\,kJ\,mol^{-1}$$
$$CH_4(g) + 2O_2(g) \rightarrow CO_2(g) + 2H_2O(g) \qquad \Delta H = -803\,kJ\,mol^{-1}$$

Show your working clearly.

2

(*b*) In the second stage, the carbon monoxide and hydrogen react to produce methanol.

$$CO(g) + 2H_2(g) \rightleftharpoons CH_3OH(g) \qquad \Delta H = -91\,kJ\,mol^{-1}$$

Circle the correct words in the table to show the changes to temperature and pressure that would favour the production of methanol.

temperature	decrease / keep the same / increase
pressure	decrease / keep the same / increase

(An additional table, if required, can be found on *Page thirty-four*.)

1

(4)

Marks

15. In an experiment to determine the enthalpy of neutralisation, dilute hydrochloric acid was added to a solution of potassium hydroxide in the apparatus shown below.

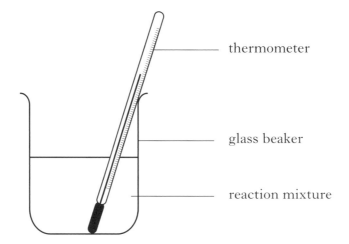

thermometer

glass beaker

reaction mixture

(a) Using the following information, calculate the enthalpy of neutralisation, in kJ mol^{-1}.

volume of 1·0 mol l^{-1} KOH = 25 cm^3

volume of 1·0 mol l^{-1} HCl = 25 cm^3

temperature change = 4·5 °C

Show your working clearly.

2

DO NOT WRITE IN THIS MARGIN

Marks

15. (continued)

(*b*) The enthalpy of neutralisation obtained using this apparatus is much smaller than the value given in data booklets.

Suggest a modification to this apparatus that would give a value closer to that given in data booklets. **1**

(*c*) What measurements would be taken to determine the temperature change? **1**

(4)

[Turn over

Marks

16. Solutions containing iodine are used to treat foot rot in sheep.

The concentration of iodine in a solution can be determined by titrating with a solution of thiosulphate ions.

$$I_2 \quad + \quad 2S_2O_3^{2-} \quad \rightarrow \quad 2I^- \quad + \quad S_4O_6^{2-}$$
thiosulphate
ions

(a) Write an ion-electron equation for the reaction of the oxidising agent in the titration.

1

(b) Three $20 \cdot 0 \, cm^3$ samples of a sheep treatment solution were titrated with $0 \cdot 10 \, mol \, l^{-1}$ thiosulphate solution.

The results are shown below.

Sample	Volume of thiosulphate/cm^3
1	$18 \cdot 60$
2	$18 \cdot 10$
3	$18 \cdot 20$

(i) Why is the volume of sodium thiosulphate used in the calculation taken to be $18 \cdot 15 \, cm^3$, although this is not the average of the three titres in the table?

1

(ii) Calculate the concentration of iodine, in $mol \, l^{-1}$, in the foot rot treatment solution.

Show your working clearly.

2

(4)

DO NOT WRITE IN THIS MARGIN

Marks

17. When a lead-acid car battery is in use, the following half reaction takes place at the negative electrode:

$$Pb(s) \quad + \quad SO_4^{2-}(aq) \quad \rightarrow \quad Pb^{2+}SO_4^{2-}(s) \quad + \quad 2e^-$$

(*a*) A lead-acid battery was used to power a car radio for 2 hours.

The radio used a current of 0·5 A.

Calculate the mass of lead, in g, converted into lead(II) sulphate.

Show your working clearly. **2**

(*b*) Complete the ion-electron equation for the reaction taking place at the other electrode.

(An additional copy of this question, if required, can be found on *Page thirty-five*.)

$$PbO_2(s) \quad + \quad SO_4^{2-}(aq) \qquad\qquad \rightarrow \quad PbSO_4(s)$$

1

(3)

[Turn over

DO NOT WRITE IN THIS MARGIN

Marks

18. Cycloalkanes are found in nature.

A representation of cyclohexane is shown below.

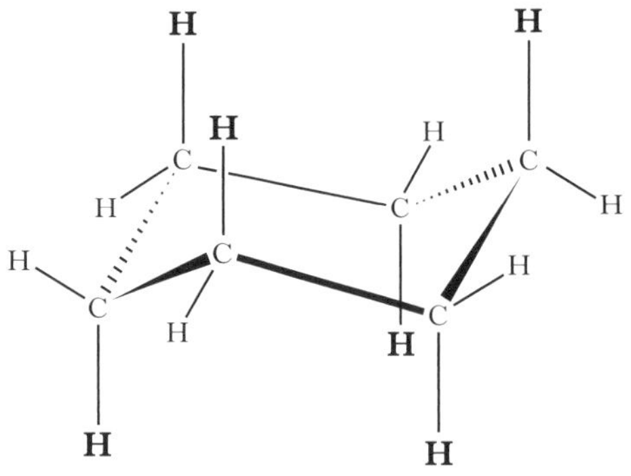

The six hydrogen atoms marked in **bold** are said to be in axial positions.

In the molecule of **1,2-**dimethylcyclohexane shown below, two methyl groups are in axial positions.

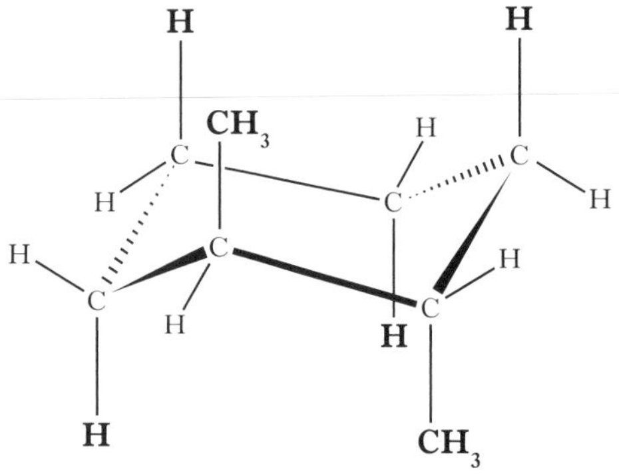

(*a*) Complete the structure below to show a molecule of **1,3-**dimethylcyclohexane in which both the methyl groups are in axial positions.

(An additional diagram, if required, can be found on *Page thirty-five*.)

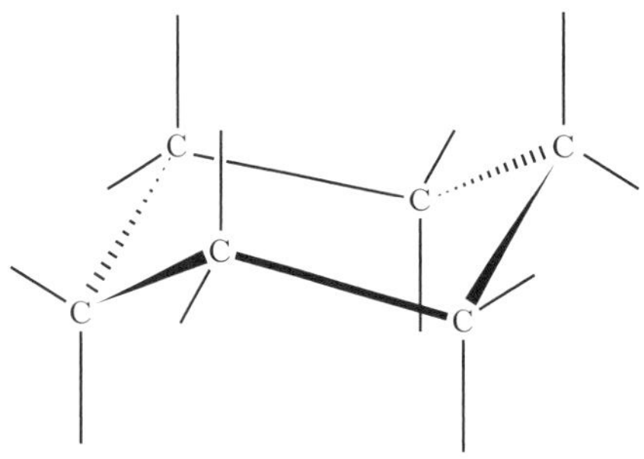

1

Marks

18. (continued)

(b) Axial groups **on the same side** of a cyclohexane ring can repel each other. The strength of the repulsion is known as the "steric strain."

The table below shows values which allow the steric strain to be calculated.

Axial groups	Steric strain/kJ mol^{-1}
H and H	0·0
H and F	0·5
H and Br	1·0
H and CH$_3$	3·8
H and (CH$_3$)$_3$C	11·4

For example:

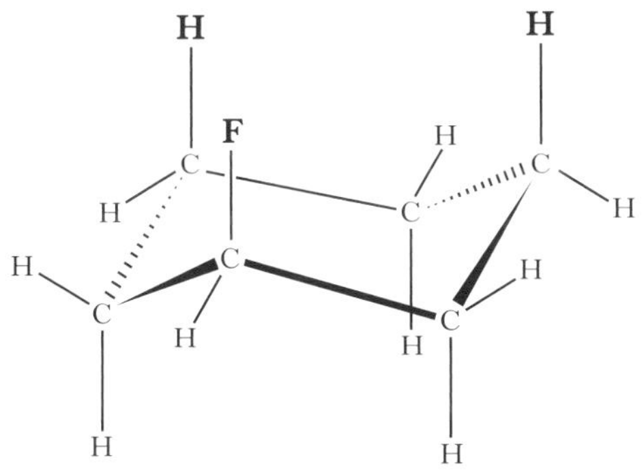

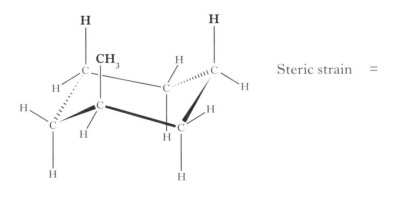

Steric strain   = 2 × (Steric strain between **H** and **F**)
=   2 × 0·5
=   1·0 kJ mol^{-1}

(i) Write a general statement, linking the size of the steric strain to the type of axial group present.

1

(ii) Calculate, in kJ mol^{-1}, the steric strain for the molecule shown below.

Steric strain   =

1

(3)

[END OF QUESTION PAPER]

ADDITIONAL DIAGRAM FOR USE IN QUESTION 5(b)

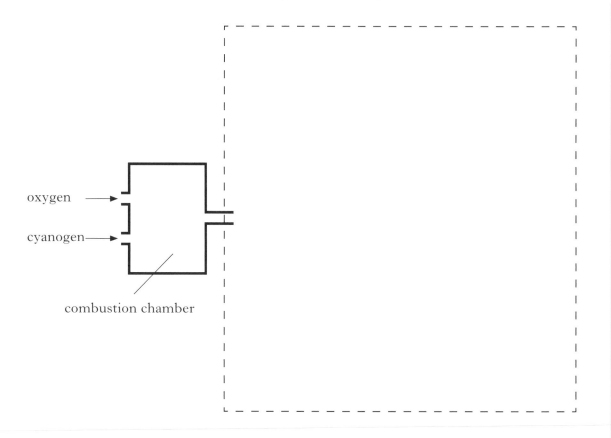

ADDITIONAL TABLE FOR USE IN QUESTION 12(a)

Mass number	
Atomic number	

ADDITIONAL TABLE FOR USE IN QUESTION 14(b)

temperature	decrease / keep the same / increase
pressure	decrease / keep the same / increase

ADDITIONAL COPY OF QUESTION 17(*b*)

17. (*b*) Complete the ion-electron equation for the reaction taking place at the other electrode.

$$PbO_2(s) \; + \; SO_4^{2-}(aq) \qquad\qquad \rightarrow \; PbSO_4(s)$$

ADDITIONAL DIAGRAM FOR USE IN QUESTION 18(*a*)

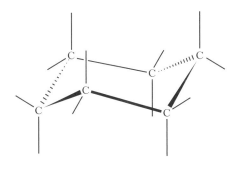

[BLANK PAGE]

HIGHER

2014

[BLANK PAGE]

FOR OFFICIAL USE

Total
Section B

X012/12/02

NATIONAL   MONDAY, 12 MAY   CHEMISTRY
QUALIFICATIONS   1.00 PM – 3.30 PM   HIGHER
2014

Fill in these boxes and read what is printed below.

Full name of centre   Town

Forename(s)   Surname

Date of birth

Day   Month   Year   Scottish candidate number   Number of seat

Reference may be made to the Chemistry Higher and Advanced Higher Data Booklet.

SECTION A—Questions 1–40 (40 marks)

Instructions for completion of **Section A** are given on page two.

For this section of the examination you must use an **HB pencil**.

SECTION B (60 marks)

1   All questions should be attempted.

2   The questions may be answered in any order but all answers are to be written in the spaces provided in this answer book, **and must be written clearly and legibly in ink**.

3   Rough work, if any should be necessary, should be written in this book and then scored through when the fair copy has been written. If further space is required, a supplementary sheet for rough work may be obtained from the Invigilator.

4   Additional space for answers will be found at the end of the book. If further space is required, supplementary sheets may be obtained from the Invigilator and should be inserted inside the **front** cover of this book.

5   The size of the space provided for an answer should not be taken as an indication of how much to write. It is not necessary to use all the space.

6   Before leaving the examination room you must give this book to the Invigilator. If you do not, you may lose all the marks for this paper.

SECTION A

Read carefully

1 Check that the answer sheet provided is for **Chemistry Higher (Section A)**.

2 For this section of the examination you must use an **HB pencil** and, where necessary, an eraser.

3 Check that the answer sheet you have been given has **your name**, **date of birth**, **SCN** (Scottish Candidate Number) and **Centre Name** printed on it.

   Do not change any of these details.

4 If any of this information is wrong, tell the Invigilator immediately.

5 If this information is correct, **print** your name and seat number in the boxes provided.

6 The answer to each question is **either** A, B, C or D. Decide what your answer is, then, using your pencil, put a horizontal line in the space provided (see sample question below).

7 There is only **one correct answer** to each question.

8 Any rough working should be done on the question paper or the rough working sheet, **not** on your answer sheet.

9 At the end of the examination, put the **answer sheet for Section A inside the front cover of your answer book**.

Sample Question

To show that the ink in a ball-pen consists of a mixture of dyes, the method of separation would be

   A chromatography

   B fractional distillation

   C fractional crystallisation

   D filtration.

The correct answer is **A**—chromatography. The answer **A** has been clearly marked in **pencil** with a horizontal line (see below).

Changing an answer

If you decide to change your answer, carefully erase your first answer and using your pencil, fill in the answer you want. The answer below has been changed to **D**.

1. Particles with the same electron arrangement are said to be isoelectronic.

   Which of the following compounds contains ions which are isoelectronic?

   A $CaCl_2$

   B   KBr

   C $MgCl_2$

   D Na_2S

2. A mixture of sodium chloride and sodium sulphate is known to contain 0·6 mol of chloride ions and 0·2 mol of sulphate ions.

   How many moles of sodium ions are present?

   A   0·4

   B   0·5

   C   0·8

   D   1·0

3. When an atom **X** of an element in Group 1 reacts to become **X**$^+$

   A   the mass number of **X** increases

   B   the charge of the nucleus increases

   C   the atomic number of **X** decreases

   D   the number of occupied energy levels decreases.

4. The reaction of copper(II) oxide with dilute sulphuric acid is an example of

   A   displacement

   B   neutralisation

   C   oxidation

   D   reduction.

5.

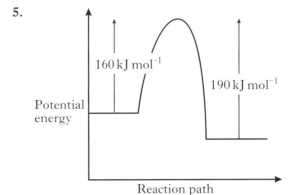

   When a catalyst is used, the activation energy of the forward reaction is reduced to $35\,kJ\,mol^{-1}$.

   What is the activation energy of the catalysed reverse reaction?

   A $30\,kJ\,mol^{-1}$

   B $35\,kJ\,mol^{-1}$

   C $65\,kJ\,mol^{-1}$

   D $190\,kJ\,mol^{-1}$

[Turn over

6. Which of the following potential energy diagrams represents the most exothermic reaction?

A

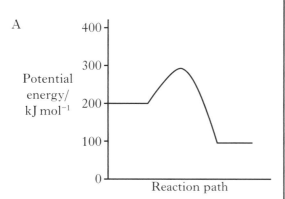

B

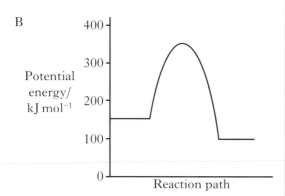

C

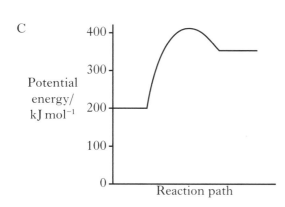

D

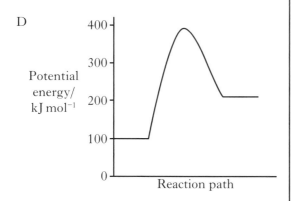

7. The following graph represents a reaction between sodium hydroxide and ethyl ethanoate.

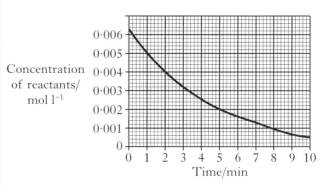

What is the average rate of the reaction over the first 5 minutes, in mol l^{-1} min^{-1}?

A　$3 \cdot 6 \times 10^{-4}$

B　$8 \cdot 4 \times 10^{-3}$

C　$8 \cdot 4 \times 10^{-4}$

D　$1 \cdot 2 \times 10^{-3}$

8. Which of the following graphs could represent the change in the rate of reaction when magnesium ribbon reacts with dilute hydrochloric acid?

A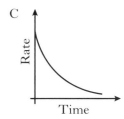

C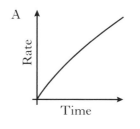

B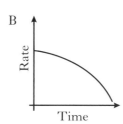

D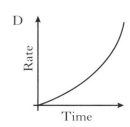

9. Graph **X** was obtained when 1 g of calcium carbonate powder reacted with excess dilute hydrochloric acid at 20 °C.

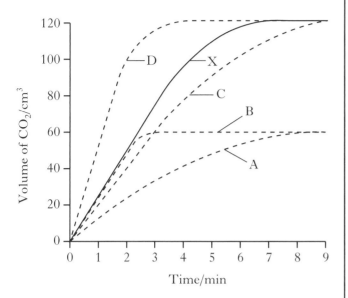

Which curve would best represent the reaction of 0·5 g lump calcium carbonate with excess of the same dilute hydrochloric acid?

10.

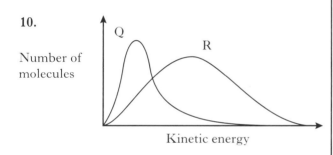

Which line in the table is correct for curves **Q** and **R** in the above graph?

	Curve Q	Curve R
A	1 mol of O_2 at 50 °C	2 mol of O_2 at 100 °C
B	1 mol of O_2 at 100 °C	2 mol of O_2 at 100 °C
C	2 mol of O_2 at 50 °C	1 mol of O_2 at 100 °C
D	2 mol of O_2 at 100 °C	1 mol of O_2 at 100 °C

11. Excess iron was added to 100 cm³ of 1·0 mol l⁻¹ copper(II) sulphate solution releasing 3·1 kJ of energy.

$$Fe(s) + CuSO_4(aq) \rightarrow Cu(s) + FeSO_4(aq)$$

What is the enthalpy change, in kJ mol⁻¹ for the above reaction?

A   −0·31

B   −3·1

C   −31

D   −310

12. For elements in Group 7 of the Periodic Table, which of the following statements is true as the group is descended?

A   The boiling point decreases.

B   The covalent radius decreases.

C   The first ionisation energy increases.

D   The strength of the van der Waals' forces increases.

13. Which line in the table is likely to be correct for the element francium?

	State at 30 °C	1st Ionisation Energy/kJ mol⁻¹
A	solid	less than 382
B	liquid	less than 382
C	solid	greater than 382
D	liquid	greater than 382

[Turn over

14. Which of the following chlorides is likely to have **least** ionic character?

   A $BeCl_2$

   B $CaCl_2$

   C   LiCl

   D   CsCl

15. Which of the following compounds has polar molecules?

   A CO_2

   B NH_3

   C CCl_4

   D CH_4

16. Which of the following are **only** found in compounds?

   A   Metallic bonds

   B   Covalent bonds

   C   Hydrogen bonds

   D   Van der Waals' forces

17. The Avogadro constant is the same as the number of

   A   atoms in 24 g of carbon

   B   molecules in 16 g of oxygen

   C   electrons in 1 g of hydrogen

   D   ions in 1 litre of sodium chloride solution, concentration $1 \, mol \, l^{-1}$.

18. Which of the following gases has the same volume as 128·2 g of sulphur dioxide?

   A   2·0 g hydrogen

   B   8·0 g helium

   C   32·0 g oxygen

   D   80·8 g of neon.

19. Which of the following statements is **not** true?

   A   Ethene and propene can be produced by cracking naphtha.

   B   Petrol can be produced by cracking naphtha.

   C   Aromatic compounds can be produced by reforming naphtha.

   D   Cycloalkanes can be produced by reforming naphtha.

20. Which of the following hydrocarbons would have the greatest tendency to auto-ignite?

   A $CH_3CH_2CH_2CH_2CH_2CH_2CH_3$

   B
$$\begin{array}{cc} CH_3 & CH_3 \\ | & | \\ CH_3CHCHCHCH_3 \\ | \\ CH_3 \end{array}$$

   C
$$\begin{array}{c} CH_2 - CH_2 \\ | \qquad\qquad CH_2 \\ CH_2 - CH_2 \end{array}$$

   D

21. MTBE and ethanol are oxygenates added to gasoline to increase the octane rating.

	Octane rating	Relative cost
MTBE	118	1·6
Ethanol	114	5·8
Gasoline	70	1·0

95 octane petrol typically contains 2·2% MTBE.

Which of the following would be true if ethanol was used instead of MTBE?

A   More ethanol would be required and the petrol would cost less.

B   More ethanol would be required and the petrol would cost more.

C   Less ethanol would be required and the petrol would cost more.

D   Less ethanol would be required and the petrol would cost less.

22. The table shows the octane numbers of four hydrocarbons.

Hydrocarbon	Octane Number
2-methylhexane	43
heptane	0
2,4-dimethylhexane	66
octane	−19

The octane number for 2,2,4-trimethylpentane is most likely to be

A   21

B   38

C   54

D   100.

23. Which of the following consumer products is **least** likely to contain esters?

A   Solvents

B   Perfumes

C   Toothpastes

D   Flavourings

24. Cured polyester resins

A   are used as textile fibres

B   are long chain molecules

C   are formed by addition polymerisation

D   have a three-dimensional structure with cross linking.

25. The structure of part of an addition polymer is shown.

$$\begin{array}{c}
\quad CH_3 \; H \quad\; CH_3 \; H \quad\; CH_3 \; H \quad\; CH_3 \; H \\
\quad\; | \quad | \quad\quad | \quad | \quad\quad | \quad | \quad\quad | \quad | \\
-C-C-C-C-C-C-C-C- \\
\quad\; | \quad | \quad\quad | \quad | \quad\quad | \quad | \quad\quad | \quad | \\
\quad H \;\; CH_3 \; H \quad H \quad H \;\; CH_3 \; H \quad H
\end{array}$$

Which pair of alkenes could be used as monomers to make this polymer?

A   But-2-ene and ethene

B   But-2-ene and propene

C   But-1-ene and ethene

D   But-1-ene and propene

[Turn over

26. Natural rubber is a polymer of isoprene

$$CH_2 = C - CH = CH_2$$
$$|$$
$$CH_3$$

The rubber formed will be

A   a condensation polymer with hydrogen bonding between the molecules

B   an addition polymer with hydrogen bonding between the molecules

C   a condensation polymer with van der Waals' forces between the molecules

D   an addition polymer with van der Waals' forces between the molecules.

27. In the formation of "hardened" fats from vegetable oils, the hydrogen

A   causes cross-linking between chains

B   causes hydrolysis to occur

C   increases the carbon chain length

D   reduces the number of carbon-carbon double bonds.

28. The chemical indicator in a breath test kit turns from orange to green when a motorist is over the legal alcohol limit.

Which of the following indicators could have been used in a breath test kit?

A   Acidified potassium dichromate

B   Benedict's solution

C   Fehling's solution

D   Tollens' reagent

29. Which of the following compounds is **not** a raw material in the chemical industry?

A   Benzene

B   Iron ore

C   Sodium chloride

D   Water

30. Which of the following is produced by a batch process?

A   Iron from iron ore

B   Aspirin from salicylic acid

C   Ammonia from nitrogen and hydrogen

D   Sulphuric acid from sulphur and oxygen.

31. $C(graphite) + O_2(g) \rightarrow CO_2(g)$ $\Delta H = -394 \, kJ \, mol^{-1}$

$C(diamond) + O_2(g) \rightarrow CO_2(g)$ $\Delta H = -395 \, kJ \, mol^{-1}$

What is the enthalpy change, in $kJ \, mol^{-1}$, for the conversion of one mole of graphite into one mole of diamond?

A -789

B -1

C $+1$

D $+789$

32. $C_2H_4(g) + 3O_2(g) \rightarrow 2CO_2(g) + 2H_2O(\ell)$ ΔH_1

$CH_3CHO(\ell) + 2\frac{1}{2}O_2(g) \rightarrow 2CO_2(g) + 2H_2O(\ell)$ ΔH_2

$2O_3(g) \rightarrow 3O_2(g)$ ΔH_3

The enthalpy change equal to $\Delta H_1 - \Delta H_2 + \frac{1}{2}\Delta H_3$ is associated with the reaction

A $C_2H_4(g) + \frac{1}{2}O_2(g) \rightarrow CH_3CHO(\ell)$

B $C_2H_4(g) + O_3(g) \rightarrow CH_3CHO(\ell) + O_2(g)$

C $C_2H_4(g) + 2O_3(g) \rightarrow CH_3CHO(\ell) + 2\frac{1}{2}O_2(g)$

D $C_2H_4(g) + 2\frac{1}{2}O_2(g) + CH_3CHO(\ell) + O_3(g) \rightarrow 4CO_2(g) + 4H_2O(\ell)$

33. Which line in the table describes dynamic equilibrium?

	Concentration of reactants and products	Forward and reverse reaction rates
A	constant	equal
B	constant	not equal
C	not constant	equal
D	not constant	not equal

34. The following reaction takes place in a blast furnace:

$$CO_2(g) + C(s) \rightleftharpoons 2CO(g) \quad \Delta H = +174\,kJ\,mol^{-1}$$

Which conditions of pressure and temperature would favour the production of carbon monoxide?

A Low pressure and low temperature

B High pressure and low temperature

C Low pressure and high temperature

D High pressure and high temperature

35. Which of the following is the same for equal volumes of $0{\cdot}1\,mol\,l^{-1}$ solutions of sodium hydroxide and ammonia?

A The pH of solution

B The mass of solute present

C The conductivity of solution

D The number of moles of hydrochloric acid needed for neutralisation

36. The graph represents the change in concentration of a reactant against time during a reversible chemical reaction.

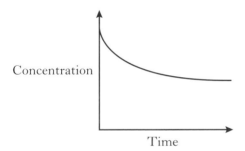

In which diagram below does the dotted line show the result of repeating the reaction using a catalyst?

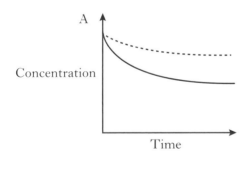

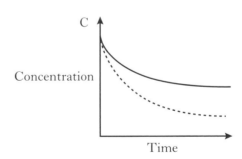

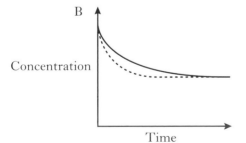

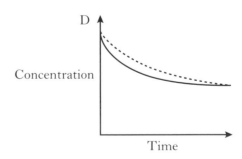

[Turn over

37. Silver jewellery discoloured by tarnish (Ag_2S) can be cleaned by placing the item in an aluminium pot containing salt solution. The reaction occurring is shown below.

$$3Ag_2S + 2Al \rightarrow 6Ag + Al_2S_3$$

Which of the following statements is true?

A   Aluminium metal is a reducing agent.

B   Silver metal is an oxidising agent.

C   Silver ions are acting as electron donors.

D   Sulphide ions are acting as electron acceptors.

38. The conductivity of pure water is low because

A   water molecules are polar

B   only a few water molecules are ionised

C   water molecules are linked by hydrogen bonds

D   there are equal numbers of hydrogen and hydroxide ions in water.

39. Limestone was added to a loch to combat the effects of acid rain. The pH of the water rose from 4 to 6.

The concentration of the $H^+(aq)$

A   increased by a factor of 2

B   increased by a factor of 100

C   decreased by a factor of 2

D   decreased by a factor of 100.

40. The diagram shows the paths of alpha, beta and gamma radiations as they pass through an electric field.

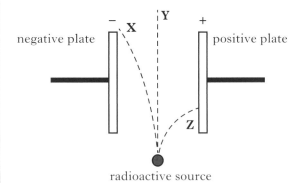

Which line in the table correctly identifies the types of radiation which follow paths **X**, **Y** and **Z**?

	Path X	Path Y	Path Z
A	alpha	beta	gamma
B	beta	gamma	alpha
C	beta	alpha	gamma
D	alpha	gamma	beta

Candidates are reminded that the answer sheet MUST be returned INSIDE the front cover of this answer book.

DO NOT WRITE IN THIS MARGIN

Marks

SECTION B

All answers must be written clearly and legibly in ink.

1. Information about four elements from the third period of the Periodic Table is shown in the table.

Element	**aluminium**	**silicon**	**phosphorus**	**sulphur**
Bonding		covalent		covalent
Structure	lattice		molecular	

(*a*) Complete the table to show the bonding and structure for each element. **2**

(*b*) What feature of the bonding in aluminium allows it to conduct electricity? **1**

(*c*) Why is there a decrease in the size of atoms across the period from aluminium to sulphur? **1**

(4)

[Turn over

Marks

2. Petrol contains branched-chain hydrocarbons, which increase the efficiency of burning.

   (a) Name the fraction from crude oil that is used to produce petrol.

   1

   (b) (i) 2,3,3-Trimethylpentane is a branched-chain hydrocarbon that is added to petrol to improve the burning efficiency.

      Draw a full structural formula for this compound.

   1

      (ii) Name one other **type** of hydrocarbon that is also added to petrol to improve the efficiency of burning.

   1

   (c) In some countries, ethanol is used as a substitute for petrol. This ethanol is produced by fermentation of glucose, using yeast enzymes.

      During the fermentation process, glucose is first converted into pyruvate. The pyruvate is then converted to ethanol in a two-step process.

$$CH_3COCOOH \xrightarrow{\text{Step 1}} CH_3CHO \xrightarrow{\text{Step 2}} CH_3CH_2OH$$
pyruvate ethanal ethanol
CO_2

      (i) Step 1 is catalysed by the enzyme, pyruvate decarboxylase.

         State **two** factors that need to be considered when choosing the best temperature at which to carry out this reaction.

   1

DO NOT WRITE IN THIS MARGIN

Marks

2. (c) continued

(ii) Why can Step 2 be described as a reduction reaction? **1**

(iii) The overall equation for the fermentation of glucose is

$C_6H_{12}O_6$	$\rightarrow$	$2C_2H_5OH$	+	$2CO_2$
mass of one mole = 180 g		mass of one mole = 46 g		

Calculate the percentage yield of ethanol if 445 g of ethanol is produced from 1·0 kg of glucose.

Show your working clearly **2**

(7)

[Turn over

Marks

3. (a) A student was asked to plan a procedure for a Prescribed Practical Activity (**PPA**).

   The student's plan is shown.

   Aim

   To find the effect of changing the concentration of iodide ions on the rate of reaction between hydrogen peroxide and an acidified solution of potassium iodide.

   Procedure

   1. Using 100 cm³ measuring cylinders, measure out 10 cm³ of sulphuric acid, 10 cm³ of sodium thiosulphate solution, 1 cm³ of starch solution and 25 cm³ of potassium iodide solution. Pour these into a dry 100 cm³ glass beaker and place the beaker on the bench.

   2. Measure out 5 cm³ of hydrogen peroxide solution and start the timer.

   3. Add the hydrogen peroxide solution to the beaker. When the blue/black colour just appears, stop the timer and record the time (in seconds).

   4. Repeat this procedure four times but each time use a different concentration of potassium iodide solution.

   (i) In step 4 of the procedure, what should be done to obtain potassium iodide solutions of different concentration? **1**

   (ii) State **two** improvements that could be made to the student's planned procedure. **1**

   (b) Collision theory can be used to explain reaction rates.

   Collision theory states that for two molecules to react, they must first collide with one another.

   State a condition necessary for the collisions to result in the formation of products. **1**

   (3)

Marks

4. (*a*) Hydrogen and fluorine can react explosively to form hydrogen fluoride gas.

The boiling point of hydrogen fluoride, HF, is much higher than the boiling point of fluorine, F_2.

H — F   F — F

boiling point: $19\cdot5\,°C$ boiling point: $-188\,°C$

Explain **fully** why the boiling point of hydrogen fluoride is much higher than the boiling point of fluorine.

In your answer you should mention the intermolecular forces involved and how they arise.

2

(*b*) Hydrogen fluoride dissolves in water to form the weak acid, hydrofluoric acid.

Dilute hydrofluoric acid reacts with sodium hydroxide solution to produce a solution of the salt sodium fluoride.

Suggest a pH for a solution of sodium fluoride.

1

(3)

DO NOT WRITE IN THIS MARGIN

Marks

5. (*a*) The polymer nylon was first synthesised by Wallace Carothers in 1935.

Part of the structure of nylon is shown.

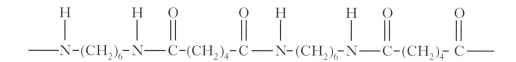

(i) What **type** of polymerisation produces nylon? **1**

(ii) During the manufacture of nylon, ethanoic acid can be added to the process to ensure that the polymer chains do not become too long.

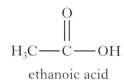

ethanoic acid

Why does adding ethanoic acid limit the polymer chain length? **1**

(*b*) Wallace Carothers' earlier research work involved the polymerisation of alkynes. This included making poly(ethyne).

Poly(ethyne) can be treated to make a polymer with an unusual property.

What is this unusual property? **1**

(3)

DO NOT
WRITE
IN THIS
MARGIN

Marks

6. Methyl orange indicator, $C_{14}H_{15}N_3SO_3$, is a weak acid.

   The equilibrium in solution is shown.

$$C_{14}H_{15}N_3SO_3(aq) \rightleftharpoons C_{14}H_{14}N_3SO_3^-(aq) + H^+(aq)$$

   red   yellow

   Explain fully the colour that would be observed when the indicator is added to dilute sodium hydroxide solution.

2

[Turn over

Marks

7. Carbon-14 is a radioactive isotope of carbon.

It is produced in the upper atmosphere when nitrogen-14 atoms are bombarded by neutrons from space. If a neutron is captured by a nitrogen-14 nucleus, a carbon-14 isotope is produced along with one other particle.

(a) Complete the nuclear equation for the formation of carbon-14.

$$^{14}_{7}\text{N} \quad + \quad ^{1}_{0}\text{n} \quad \rightarrow \quad ^{14}_{6}\text{C} \quad +$$

1

(b) Carbon-14 decays by beta emission. Why does the atomic number increase by one unit when a carbon-14 nucleus decays?

1

(c) Carbon dating can be used to estimate the age of wood found in archaeological sites.

The decay curve shows the decrease in the percentage of carbon-14 against the number of half-lives.

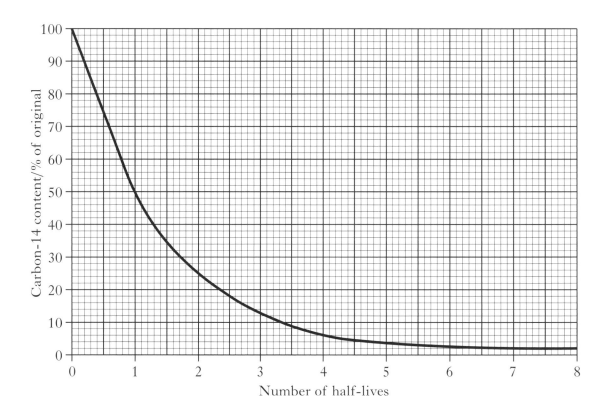

DO NOT WRITE IN THIS MARGIN

Marks

7. (*c*) continued

(i) A piece of wood was found to contain 5% of the original carbon-14 content. The half-life of carbon-14 is 5700 years. Calculate the age of the wood, in years. **1**

(ii) Suggest a reason why carbon-14 is unsuitable for dating samples that are more than 50 000 years old. **1**

(4)

[Turn over

Marks

8. Self-heating cans may be used to warm drinks such as coffee.

   When the button on the can is pushed, a seal is broken allowing water and calcium oxide to mix and react.

   The reaction produces solid calcium hydroxide and releases heat.

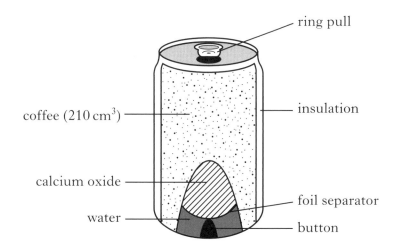

   The equation for this reaction is:

   $$CaO(s) + H_2O(\ell) \rightarrow Ca(OH)_2(s) \qquad \Delta H = -65 \text{ kJ mol}^{-1}$$

   (a) Calculate the mass, in grams, of calcium oxide required to raise the temperature of 210 cm³ of coffee from 20 °C to 70 °C.

   Show your working clearly.

2

Marks

8. continued

(*b*) If more water is used the calcium hydroxide is produced as a solution instead of as a solid.

The equation for the reaction is:

$$CaO(s) + H_2O(\ell) \rightarrow Ca(OH)_2(aq)$$

Using the following data, calculate the enthalpy change, in kJ mol^{-1}, for this reaction.

$$Ca(s) + \tfrac{1}{2}O_2(g) \rightarrow CaO(s) \qquad \Delta H = -635 \, kJ \, mol^{-1}$$

$$H_2(g) + \tfrac{1}{2}O_2(g) \rightarrow H_2O(\ell) \qquad \Delta H = -286 \, kJ \, mol^{-1}$$

$$Ca(s) + O_2(g) + H_2(g) \rightarrow Ca(OH)_2(s) \qquad \Delta H = -986 \, kJ \, mol^{-1}$$

$$Ca(OH)_2(s) \rightarrow Ca(OH)_2(aq) \qquad \Delta H = -82 \, kJ \, mol^{-1}$$

Show your working clearly.

2

(4)

[Turn over

Marks

9. (*a*) An ester can be prepared from a mixture of ethanol and methanoic acid.

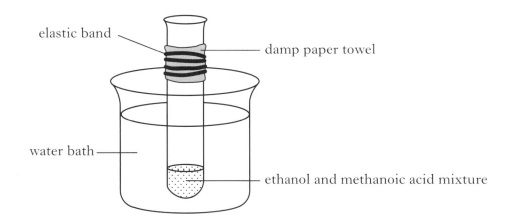

elastic band — damp paper towel

water bath — ethanol and methanoic acid mixture

   (i) Name another substance that should be added to the mixture.

1

   (ii) Why should the reaction mixture be heated using a water bath and not a Bunsen burner?

1

   (iii) Draw a structural formula for the ester that would be produced in this reaction.

1

(*b*) Sodium methanoate is a food additive (E237). It can be prepared by reacting chloroform, $CHCl_3$, with sodium hydroxide.

$$CHCl_3(\ell) + NaOH(aq) \rightarrow HCOONa(aq) + NaCl(aq) + H_2O(\ell)$$

Balance this equation.

1

(4)

Marks

10. Ammonia is produced by the Haber process from nitrogen and hydrogen in the presence of an iron catalyst.

$$N_2(g) + 3H_2(g) \rightleftharpoons 2NH_3(g) \qquad \Delta H = -92 \, kJ \, mol^{-1}$$

(a) Name the **type** of catalysis taking place in the Haber process.

1

(b) The source of hydrogen gas for the Haber process is methane. Methane is steam-reformed to produce carbon monoxide and hydrogen.

$$CH_4(g) + H_2O(g) \rightarrow CO(g) + 3H_2(g)$$

The carbon monoxide is further reacted with steam to produce carbon dioxide and hydrogen.

$$CO(g) + H_2O(g) \rightarrow CO_2(g) + H_2(g)$$

Write the overall equation for the reforming process.

1

(c) An alternative source of hydrogen involves the electrolysis of water.

(i) Calculate the volume, in litres, of hydrogen produced when a current of 200 A is passed through acidified water for 30 minutes.

(Take the volume of 1 mole of hydrogen to be 24 litres)

The ion-electron equation for the process is:

$$2H^+(aq) + 2e^- \rightarrow H_2(g)$$

2

(ii) **In terms of the products**, suggest an advantage of producing hydrogen by electrolysis of water rather than by steam reforming of methane.

1

(5)

Marks

11. The active chemical in CS spray was developed by two chemists, Corson and Stoughton, after whom it is named.

(*a*) The active "CS" chemical has the structure shown.

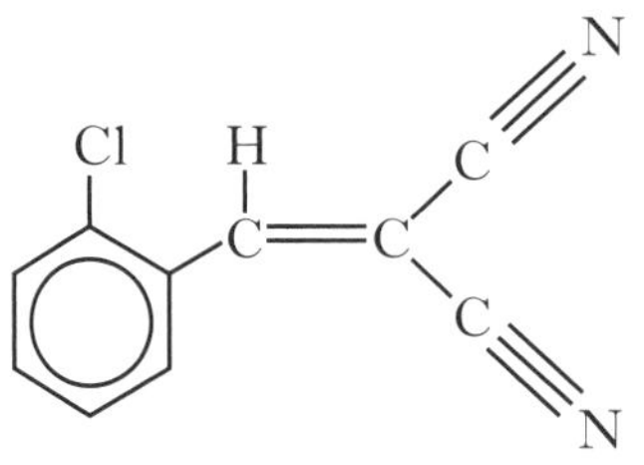

The molecular formula for this compound can be written as $C_wH_xN_yCl_z$.

Give the values for w, x, y and z.

w =   x =   y =   z =   **1**

(*b*) The solvent used in CS spray is commonly known as MiBK and has the structure shown.

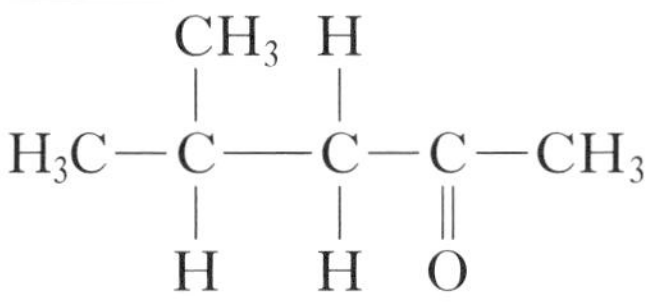

Give the systematic name for this solvent.

1

(*c*) The propellant gas in a CS canister is nitrogen.

Previously, CFCs were widely used as the propellant gases for many sprays.

Why was the use of CFCs discontinued?

1

Marks

11. (continued)

(d) The MiBK solvent is manufactured from propanone as shown in the following reaction sequence.

Step 1 Two molecules of propanone react

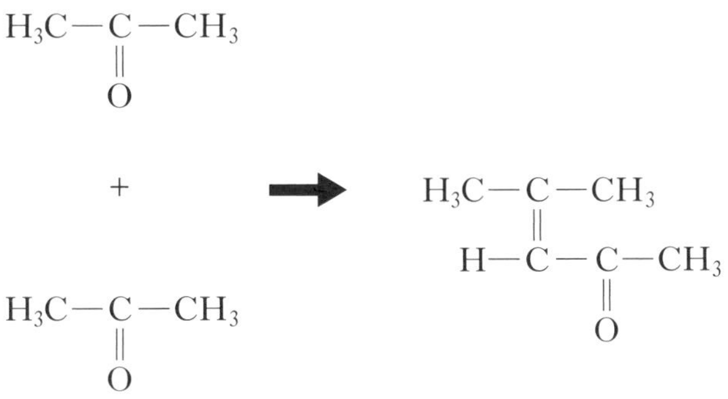

$$H_3C-C-CH_3$$
$$\overset{\|}{O}$$

+   ⟶ $$H_3C-C-CH_3$$
                    $$\overset{\|}{H-C}-C-CH_3$$
$$H_3C-C-CH_3$$ $$\overset{\|}{O}$$
$$\overset{\|}{O}$$

In this reaction the carbon to carbon double bond forms between the carbonyl group of one molecule and the α-carbon of the second molecule (the α-carbon is the carbon adjacent to the carbonyl group).

Step 2

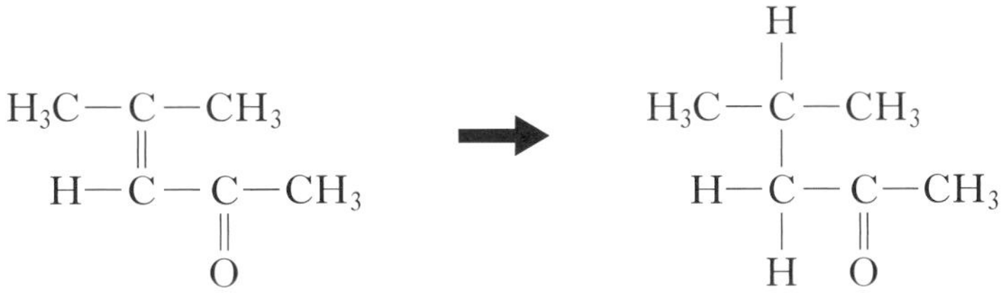

$$H_3C-C-CH_3$$
$$\overset{\|}{H-C}-C-CH_3$$
$$\overset{\|}{O}$$

⟶

$$\overset{H}{\underset{|}{H_3C-C-CH_3}}$$
$$\overset{|}{H-C}-C-CH_3$$
$$\overset{|}{H}\ \overset{\|}{O}$$

MiBK

(i) What is the name for the type of addition reaction taking place in **Step 2**?

1

(ii) Draw the product formed at the end of **Step 1** when a molecule of propanone reacts with a molecule of methanal.

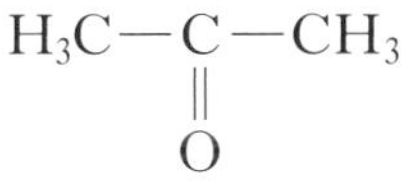

$$H_3C-C-CH_3$$ $$H-C-H$$
$$\overset{\|}{O}$$ $$\overset{\|}{O}$$
propanone 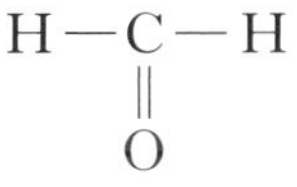   methanal

1

(5)

DO NOT WRITE IN THIS MARGIN

Marks

12. Proteins are made from monomers called amino acids.

Human hair is composed of long strands of a protein called keratin.

(*a*) What **type** of protein is keratin? **1**

(*b*) Part of the structure of a keratin molecule is shown.

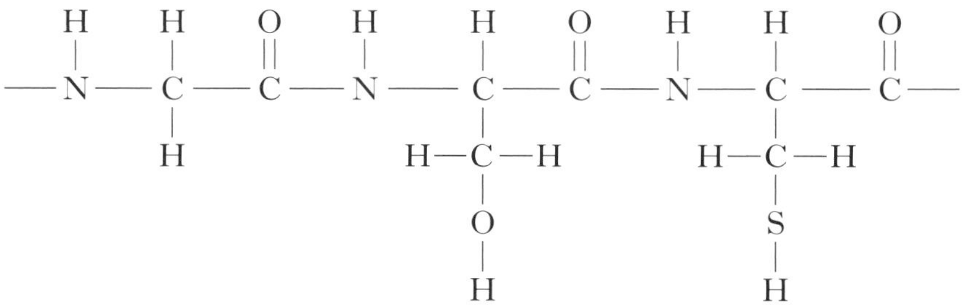

Circle a peptide link in the structure. **1**

(*c*) Hair products contain a large variety of different chemicals.

Chemicals called hydantoins are used as preservatives in shampoos to kill any bacteria.

A typical hydantoin is shown.

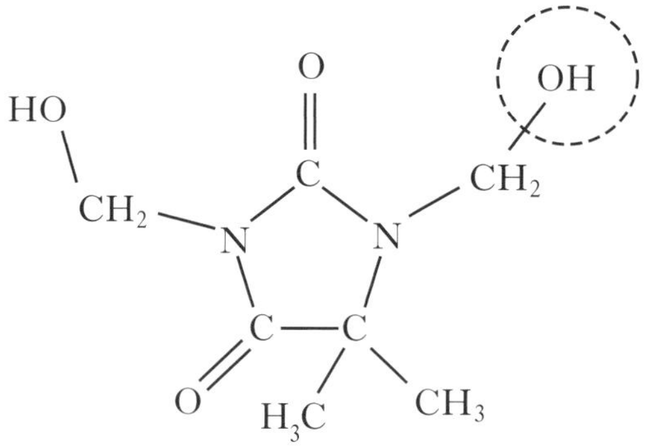

Name the functional group circled. **1**

Marks

12. **(continued)**

(*d*) Some hair conditioners contain the fatty acid behenic acid, $CH_3(CH_2)_{19}CH_2COOH$.

Behenic acid is produced by hydrolysing the edible oil, ben oil.

(i) Name the compound, other than fatty acids, which is produced by hydrolysing ben oil.

1

(ii) When conditioner containing behenic acid is applied to hair, the behenic acid molecules make intermolecular hydrogen bonds to the keratin protein molecules.

On the diagram below use a dotted line to show **one** hydrogen bond that could be made between a behenic acid molecule and the keratin.

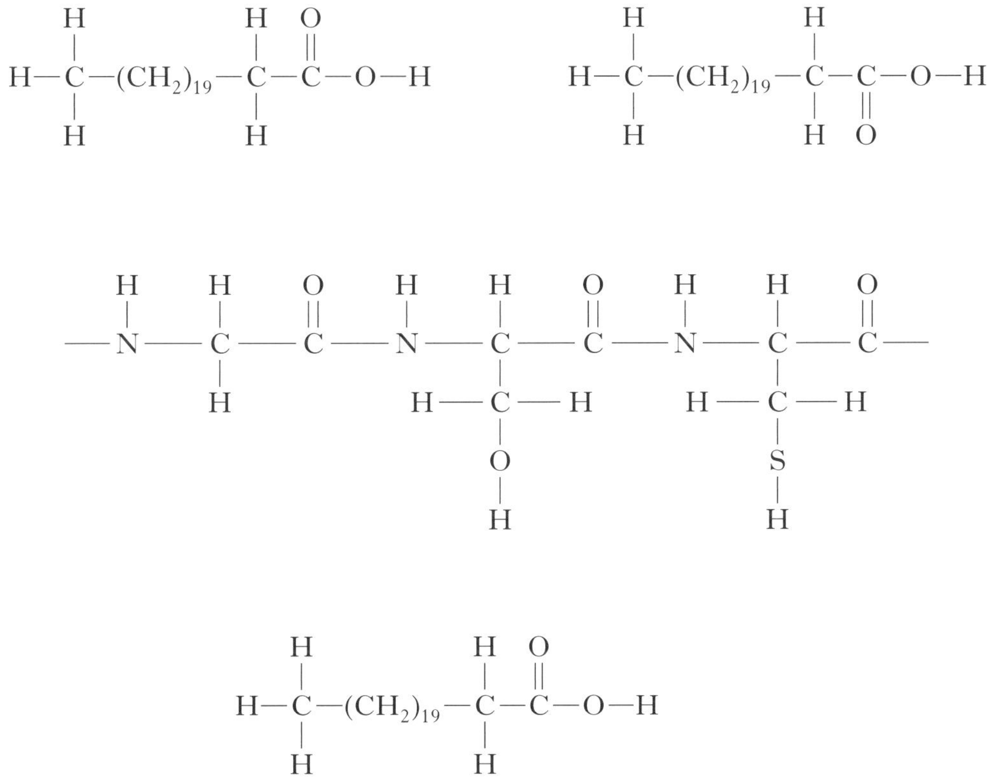

1

(5)

[**Turn over**

Marks

13. The vitamin C content in a fruit drink can be determined by titrating it with iodine.

$$C_6H_8O_6(aq) + I_2(aq) \rightarrow C_6H_6O_6(aq) + 2H^+(aq) + 2I^-(aq)$$
vitamin C

To determine the vitamin C content in a 1·0 litre carton of orange juice, three separate $20\,cm^3$ samples of the juice were titrated with a $0·00125\ mol\,l^{-1}$ iodine solution. Starch indicator was used to determine the endpoint.

(a) The iodine solution is dark brown in colour which makes reading the scale on the burette difficult.

How can this difficulty be overcome?

1

(b) What colour indicates the endpoint of the titration?

1

(c) The following results were obtained from titration of the three $20\,cm^3$ samples of orange juice.

Titration	Volume of $0·00125\ mol\,l^{-1}$ iodine solution used/cm^3
1	26·3
2	25·5
3	25·3

Average volume of iodine solution used = $25·4\ cm^3$

Suggest why the volume of iodine solution added in the first titration was higher than that added in the other two titrations.

1

Marks

13. **(continued)**

(d) $C_6H_8O_6(aq) \quad + \quad I_2(aq) \quad \rightarrow \quad C_6H_6O_6(aq) \quad + \quad 2H^+(aq) \quad + \quad 2I^-(aq)$
vitamin C
1 mol   1 mol

Calculate the mass, in grams, of vitamin C, in the 1·0 litre carton of orange juice.

(mass of 1 mole vitamin C = 176 g)

Show your working clearly.

2

(5)

[Turn over

Marks

14. A chemical explosion is the result of a very rapid reaction that generates a large quantity of heat energy and, usually, a large quantity of gas.

    (a) The explosive RDX, $C_3H_6N_6O_6$, is used in the controlled demolition of disused buildings.

    During the explosion it decomposes as shown.

    $$C_3H_6N_6O_6(s) \quad \rightarrow \quad 3CO(g) \quad + \quad 3H_2O(g) \quad + \quad 3N_2(g)$$

    Calculate the volume of gas released when $1\cdot0\,g$ of RDX decomposes.

    (Take the molar volume to be 24 litres mol^{-1})

    Show your working clearly

2

DO NOT WRITE IN THIS MARGIN

14. (continued)

Marks

(b) The products formed when an explosive substance decomposes can be predicted by applying the Kistiakowsky-Wilson rules. These rules use the number of oxygen atoms in the molecular formula to predict the products.

In the example below these rules are applied to the decomposition of the explosive RDX, $C_3H_6N_6O_6$

Rule Number	Rule	Atoms available in $C_3H_6N_6O_6$	Apply Rule to show products
1	Using oxygen atoms from the formula convert any carbon atoms in the formula to carbon monoxide	$3 \times C$	$3CO$ formed
2	If any oxygen atoms remain convert H atoms in the formula to water	$3 \times O$ remain	$3H_2O$ formed
3	If any oxygen atoms still remain then convert CO formed to CO_2	No more oxygen left	No CO_2 formed
4	Convert any nitrogen atoms in the formula to N_2	$6 \times N$	$3N_2$ formed

Decomposition equation:

$$C_3H_6N_6O_6(s) \rightarrow 3CO(g) + 3H_2O(g) + 3N_2(g)$$

By applying the same set of rules, complete the equation for the decomposition of the explosive PETN, $C_5H_8N_4O_{12}$.

$C_5H_8N_4O_{12}(s) \rightarrow$

1

(3)

[Turn over

Marks

15. In solution, amino acid molecules can form zwitterions when a hydrogen ion moves from the carboxyl group onto the amino group.

For example,

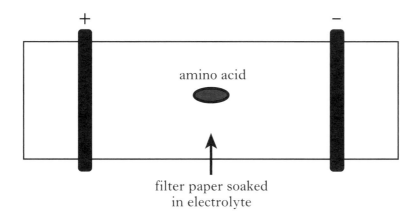

glycine   glycine zwitterion

(a) Draw the zwitterion produced by the amino acid serine.

$$HO-\overset{\overset{\displaystyle O}{\|}}{C}-\overset{\overset{\displaystyle H}{|}}{\underset{\underset{\displaystyle NH_2}{|}}{C}}-\overset{\overset{\displaystyle H}{|}}{\underset{\underset{\displaystyle H}{|}}{C}}-OH \longrightarrow$$

serine   serine zwitterion   1

(b) Having both positive and negative charges, an amino acid zwitterion is electrically neutral and will not move if placed between oppositely charged electrodes.

+   –

amino acid

filter paper soaked
in electrolyte

Zwitterions only exist at a specific pH, which varies for each amino acid. This specific pH is called the isoelectric point (IEP). At pH's higher than the IEP, amino acids in solution exist as negative ions and move towards the positive electrode. At pH's lower than the IEP, they exist as positive ions and move towards the negative electrode.

Marks

15. (b) continued

For example,

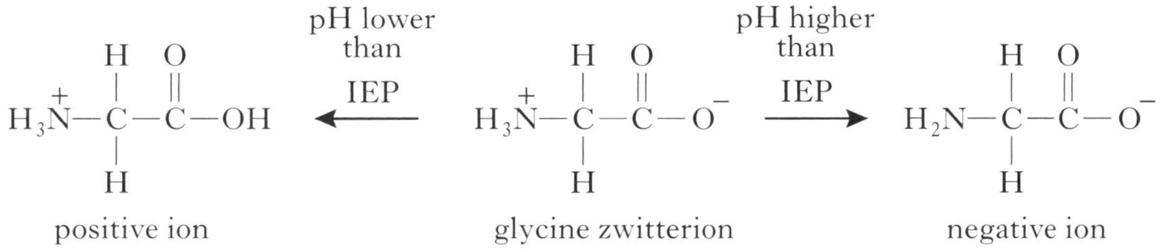

$$H_3\overset{+}{N}-\underset{\underset{H}{|}}{\overset{\overset{H}{|}}{C}}-\overset{\overset{O}{\|}}{C}-OH \quad \overset{\text{pH lower}}{\underset{\text{IEP}}{\overset{\text{than}}{\longleftarrow}}} \quad H_3\overset{+}{N}-\underset{\underset{H}{|}}{\overset{\overset{H}{|}}{C}}-\overset{\overset{O}{\|}}{C}-O^- \quad \overset{\text{pH higher}}{\underset{\text{IEP}}{\overset{\text{than}}{\longrightarrow}}} \quad H_2N-\underset{\underset{H}{|}}{\overset{\overset{H}{|}}{C}}-\overset{\overset{O}{\|}}{C}-O^-$$

positive ion   glycine zwitterion   negative ion

Lysine is an amino acid with an IEP of pH 9·7.

Explain clearly whether lysine would be attracted to the positive or negative electrode in a $1·0 \times 10^{-5}$ mol l^{-1} sodium hydroxide solution.

2

(3)

[*END OF QUESTION PAPER*]

ADDITIONAL SPACE FOR ANSWERS

ADDITIONAL SPACE FOR ANSWERS

[BLANK PAGE]

HIGHER | ANSWER SECTION

CHEMISTRY HIGHER
2010

SECTION A

1. B	11. B	21. C	31. B
2. C	12. A	22. C	32. A
3. C	13. B	23. A	33. B
4. B	14. D	24. C	34. C
5. B	15. D	25. A	35. B
6. D	16. C	26. D	36. C
7. C	17. D	27. D	37. B
8. B	18. A	28. C	38. D
9. A	19. A	29. D	39. D
10. D	20. B	30. D	40. A

SECTION B

1. lithium   metallic (or metal)
   boron   covalent   network or lattice
   nitrogen   (discrete) molecular (or molecule) or diatomic

2. (a) (i) 8
   (ii)

   $$-\overset{\overset{\textstyle O}{\|}}{C}-O-\overset{\overset{\textstyle O}{\|}}{C}-\overset{\overset{\textstyle H}{|}}{N}-O-\underset{\underset{\textstyle OH}{}}{\overset{\overset{\textstyle H}{|}}{N}}-$$

   (b) dissolves (or soluble) in water

3. (a) (i) rate of forward reaction equals rate of reverse reaction or concentration of reactants and products remain constant
   (ii) decreases (or reduces or gets smaller or diminishes or lowers)

   (b) no. of moles $= \dfrac{0\cdot010}{32} = 3\cdot125 \times 10^{-4}$

4. (a) they react with the oxygen (or are oxidised) or burn or react to form CO_2 or CO

   (b) Q = I t = 50 000 × 20 × 60 = 6×10^7 C
   Al 3 × 96 500 C ⟷ 1 mol
   6×10^7 C ⟷ $\dfrac{6 \times 10^7 \times 27}{3 \times 96\,500} = 5596$ g

5. (a) (i) concentration of   volume of reactants
   reactants

   (or permanganate or   (or permanganate
   oxalic acid)   or oxalic acid)

   (ii) colour change is too slow (or too gradual or takes a long time) or colour change is indistinct

   (b)

   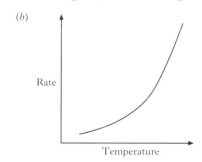

6. (a) $^{11}C \quad \rightarrow \quad ^{11}B \quad + \quad ^{\;0}_{\;1}e$

   (b) 3 half-lives   half-life $= \dfrac{60}{3} = 20$ minutes

   (c) ^{11}C
   more ^{11}C atoms or more radioactive atoms or greater mass of ^{11}C or ^{11}C has no other elements

7. (a) intermolecular attractions (or forces) or attractions between molecules

   (b)

   $$CH_3-\overset{\overset{\textstyle OH}{|}}{\underset{\underset{\textstyle CH_3}{|}}{C}}-\overset{\overset{\textstyle O}{\text{//}}}{\underset{\underset{\textstyle OH}{\backslash}}{C}}$$

8. (a) (i) a reactant from which other chemicals can be made (or synthesised or produced or obtained or derived) or product of one reaction becomes the reactant of another
   (ii) addition (or additional)
   (iii) sodium chloride
   (iv) fats and oils are renewable (or will not run out or are unlimited) or propene is obtained from a finite source

   (b) $2C_3H_8O_3 \quad \rightarrow \quad 3CO_2 \quad + \quad 3CH_4 \quad + \quad 2H_2$

   (c) $3C \; + 3O_2 \; \rightarrow 3CO_2 \quad -394 \times 3 \; = -1182$ kJ
   $4H_2 \; + 2O_2 \; \rightarrow 4H_2O \quad -286 \times 4 \; = -1144$ kJ
   $3CO_2 + 4H_2O \rightarrow C_3H_8O_3 + 7/2\,O_2 \; = +1654$ kJ
                addition $= -672$ kJ mol^{-1}

9. (a) carbon, oxygen, nitrogen and hydrogen

   (b) count the number of (oxygen or gas) bubbles produced in a given time or measure the volume of gas produced in a given time or measure height of bubbles (or foam) produced in a given time or find rate of gas production

   (c) increasing temperature can denature the enzyme or idea of optimum temperature

10. (a) <u>for drying</u>, entry delivery tubes must be below surface of concentrated sulphuric acid and exit tube must be above

    <u>for collection</u>, apparatus must be workable and 'cooler' labelled, eg use of an ice/water bath

    (b) 1 mol SO_2 →   1 mol SO_3
    64·1g   →   80·1g
    51·2 tonnes → $\dfrac{51\cdot2 \times 80\cdot1}{64\cdot1} = 64\cdot0$ tonnes

    % yield $= \dfrac{\text{actual}}{\text{theoretical}} \times 100 = \dfrac{43\cdot2}{64\cdot0} \times 100 = 67\cdot5\%$
    or
    moles of $SO_2 = \dfrac{51\cdot2}{64\cdot1} = 0\cdot799$

    moles of $SO_3 = \dfrac{43\cdot2}{80\cdot1} = 0\cdot539$

    % yield $= \dfrac{\text{actual}}{\text{theoretical}} \times 100 = \dfrac{0\cdot539}{0\cdot799} \times 100 = 67\cdot5\%$

11. (a) (i) outer electron is further away from the nucleus or greater number of electron shells
    and
    (increased) shielding (or screening) by the inner electrons or decreased nuclear attraction due to inner election shells
    (ii) $3\cdot94 \times 10^{-21} \times 6 \times 10^{23} = 2371\cdot9$ kJmol^{-1}

    (b) $Cl(g) + e^- \rightarrow Cl^-(g)$

12. (a) moles of LiOH = $0.1 \times 0.4 = 0.04$

moles of $CO_2 = \dfrac{0.24}{34} = 0.01$

0.02 mol of LiOH reacts with 0.01 mol of CO_2

excess LiOH = 0.02

(b) 13

(c) two points related to weak acid equilibrium
two points related to water equilibrium
or
salt of a weak acid and a strong base for one mark

13. (a) (i)

$$CH_3 \underset{\underset{CH_3}{|}}{\overset{\overset{CH_3}{|}}{C}} \underset{\underset{H}{|}}{\overset{\overset{CH_3}{|}}{C}} CH_3$$

(ii) all have branched-chains (or branches)

(b) (i) more complete combustion (or less incomplete combustion or less CO) or higher octane rating or burns more smoothly or prevents knocking (or auto-ignition), or carbon burns more cleanly or reduces the oxygen (or air) required for combustion
(ii) any correct ether isomer

(c) cyclohexane or any correct cyclic isomer

14. (a) (i) 2. measure the temperature (of the water)
4. measure the highest temperature reached by the solution
(ii) to reduce (or prevent) heat loss to the surroundings or to keep heat in or less energy lost (or to conserve energy)
(iii) 1 mol KOH = 56.1 g

$1.2\,g \leftrightarrow 1.08\,kJ$

$56.1 \leftrightarrow \dfrac{1.08 \times 56.1}{1.2} = -50.49\,kJ\,mol^{-1}$

(b) enthalpy change is for the formation of **one** mole of water or equivalent

15. (a) **x** is O-H   **y** is C-H

(b) (i) condensation or esterification
(ii) 2 peaks only: at 1705-1800 and 2800-3000

16. (a) *Any two from:*
flask should be swirled
read burette at eye level
white tile under flask
add drop-wise (near end-point)
no air bubble in burette
use an indicator to give a sharp colour change
rinse with solutions being used
titrate slowly
remove funnel from burette
put a piece of white paper behind burette
stir constantly, etc.

(b) (i) no. of moles of MnO_4^- (aq) = $21.6 \times 1.50 \times 10^{-5}$
$= 3.24 \times 10^{-4}$

mole ratio 2:5

no. of moles of $NO_2^- = 8.1 \times 10^{-4}$

concentration $= \dfrac{8.1 \times 10^{-4}}{0.025} = 3.24 \times 10^{-2}\,mol^{-1}$

(ii) $NO_2^-(aq) + H_2O(l) \rightarrow NO_3^-(aq) + 2H^+(aq) + 2e^-$

SECTION A

1. D	11. C	21. B	31. B
2. A	12. C	22. A	32. D
3. A	13. A	23. C	33. D
4. D	14. D	24. B	34. A
5. B	15. C	25. D	35. B
6. D	16. D	26. D	36. C
7. B	17. D	27. B	37. C
8. C	18. A	28. B	38. A
9. C	19. C	29. D	39. B
10. A	20. B	30. C	40. A

SECTION B

1. (a) homogeneous

(b) (i) answer 0.0015
(ii) new line should start **at same point as original** and should have a **steeper gradient**

2. (a) (i) more protons or increasing nuclear charge
(ii) $Cl(g) \rightarrow Cl^+(g) + e^-$

(b) argon does not form (covalent) bonds
or
no electrons involved in bonding

3. (a) covalent bonds not being broken
or
intermolecular bonds that are breaking

(b) formula refers to the ratio of $Mg^{2+}:Cl^-$ ions (in lattice) (or alternative wording ie in the lattice there are twice as many chloride ions as magnesium ions)
or
Mg^{2+} ions surrounded by > 2 Cl^- ions
or
Cl^- surrounded by >1 Mg^{2+}

4. (a) 2,2,4-trimethylpentane

(b) it has more volatile (compounds)/vaporise more easily
or
(hydrocarbons) boil more easily/lower boiling point
or
more short chain compounds/lower GFM/more butane
or
less viscous

(c) *Any four from the following for a maximum of two marks:*
½ mark for safe heating method (no flame)/water bath
½ mark for condenser of some type
½ mark for methanol and stearic acid or "reactants"
½ mark for (concentrated) sulphuric acid in test tube
½ mark for pouring the mixture into a carbonate solution or solid carbonate added after esterification

5. (a) 1 mole $Ca(OCl)_2 \rightarrow$ 2 moles Cl_2

$143g \rightarrow 48$ litres

$\dfrac{0.096}{48} \times 143$

$= \underline{\textbf{0.286g or 0.29g}}$
or
moles of Cl_2 $\dfrac{0.096}{24} = 0.004$

moles of $Ca(OCl)_2$ $\dfrac{0.004}{2} = 0.002$

mass of $Ca(OCl)2 = 0.002 \times \underline{143g}$

$= \underline{\textbf{0.286g}}$

(b)

6. (a) $CH_3-CH_2-\underset{\underset{OH}{|}}{CH}-CH_3$

(b) ½ mark for triethanol amine has <u>hydrogen bonds</u> (between the molecules)

½ mark for triisopropyl amine molecules has van der Waals/or permanent dipole/permanent dipole attractions or doesn't have H-bonds

½ mark for H-bonds strong(er) (than the dipole/dipole)

½ mark for more energy/higher temp required (to overcome/break intermolecular forces)

7. (a) $C_8H_9NO_2$

(b) amino acids

(c) 0.0225 or 0.022 or 0.023
(can be rounded to 0.02 if working given)

8. (a)

or

(b) **either**
1 mole glycerol → 1 mole ethane-1,2-diol
    92g   →   62g
27.6kg   →   18.6kg

% yield $= \dfrac{13.4}{18.6} \times 100$

% yield = 72 %

or

moles of glycerol $= \dfrac{27600}{92}$

moles of glycerol $= 300$

actual moles ethane-1,2-diol $= \dfrac{13400}{62}$

actual moles of ethane-1,2-diol $= 216.13$

% yield $= \dfrac{216.13}{300} \times 100$

% yield = 72 %

9. (a) palm oil has lower degree of unsaturated/palm oil less unsaturated/palm oil more saturated/palm oil contains more saturates/fewer double bounds
or
molecules in palm oil can pack more closely together

(b) polyunsaturated

(c) soap/emulsifying agent/detergent/washing/cleaning

10. (a) (i) $O_3 + 2KI + H_2O \rightarrow I_2 + O_2 + 2KOH$
(ii) purple or blue/black or black or blue

(b) ½ mark for power supply/battery/lab pack
½ mark for (dilute sulphuric) acid labelled
½ mark for method for collecting O_3 which would work at positive electrode

(c) ½ mark for O_3 collected
(i) acidified dichromate (solution)
(ii)

or

11. (a) partially ionised/not completely dissociated

(b) (i) contains more H^+ ions/higher concentration of H^+ ions
(ii) because it is diprotic/dibasic/has two hydrogens
or balanced equations for the reactions
or sulphurous acid has more hydrogens
or sulphurous acid has a high power of hydrogen
(iii) pH = 13

12. (a) neutron to proton ratio (is unstable)
or proton to neutron ratio (is unstable)
or they have too many/few neutrons

(b) $^{131}_{53}I \rightarrow {}^{131}_{54}Xe + {}^{0}_{-1}e$

$^{131}_{53}I \rightarrow {}^{131}_{54}Xe + {}^{0}_{-1}e^-$

$^{131}I \rightarrow {}^{131}Xe + e^-$

$^{131}I \rightarrow {}^{131}Xe + e$

$^{131}I \rightarrow {}^{131}Xe + \beta$

(c) (i) 8 days
(ii) ½ mark for correct data from graph 70
½ mark for conversion to mole ($\div 131$) 5.343×10^{-13}
½ mark for use of 6.02×10^{23}
½ mark for answer 3.22×10^{11} (ions)

13. (a) on addition of NaOH(s):
• OH^- react with H^+
• concentration of H^+ decreases
• equilibrium position to shift to the left
• CrO_4^{2-} ion concentration increases
• Solution becomes more yellow/less orange

(b) (i) • mention of washings/rinsings
• make the (standard) flask up to the mark with water/add water until desired volume reached
(ii) **either**
moles $FeSO_4$ $0.02 \times 0.0274 = 0.000548$

moles of CrO_4^{2-} $\dfrac{0.000548}{3} = 0.000183$

concentration of CrO_4^{2-} $\dfrac{0.000183}{0.050}$

$= 0.00365$ or 0.004(mol I^{-1})

or

candidates may use a "titration" formula of which an example is shown below.

$$\frac{C_1 V_1}{b_1} = \frac{C_2 V_2}{b_2}$$

$$\frac{C1 \times 50.0}{1} = \frac{0.0200 \times 27.4}{3}$$

$$C_1 = \frac{0.0200 \times 27.4}{3 \times 50.0}$$

concentration of CrO_4^{2-} = 0.00365 (mol I^{-1})

14. (a) answer within range −2640 to −2690

(b) E = mcΔT = 0.2 × 4.18 × 40 = 33.44

74g gives 33.44 × 74 = 2475/2477 kJ

enthalpy of comb. = −2475/−2477

(c) reversing given equation +354
using enthalpy of combustion of C−5 × 394 or −1970
using enthalpy of combustion of H_2 −6 × 286 or −1716

addition −3332

15. (a) precipitation

(b) compound Z is water, **or** H_2O **or** steam **or** hydrogen oxide

(c) (i) the chlorine gas produced during the electrolysis of cerium chloride can be recycled/ reused (back into stage 4)

or

a substance may be added to reduce the temperature at which $CeCl_3$ melts

or

$CeCl_3$ can be electrolysed in solution (to avoid heating costs for $CeCl_3$(l) electrolysis)

(ii) Q = It

Q = 4000 × 10 × 60

Q = 2400000C

Ce^{3+} + $3e^-$ → Ce

3 × 96500C → 140.1g

2400000 C → 1161.45g or 1.16 kg

16. (a) 3

(b) 0.204°C

SECTION A

1. D	11. D	21. A	31. B
2. C	12. B	22. B	32. D
3. D	13. B	23. B	33. A
4. C	14. C	24. A	34. C
5. B	15. D	25. C	35. B
6. D	16. B	26. D	36. D
7. C	17. A	27. D	37. D
8. B	18. A	28. C	38. A
9. A	19. C	29. B	39. B
10. C	20. A	30. A	40. C

SECTION B

1. (a) boron or carbon or B or C or graphite or diamond

(b) number of protons <u>increases</u>
or <u>increased</u> atomic number
or <u>greater</u> nuclear/positive charge
or <u>greater</u> pull on (outer) electrons

2. (a) to prevent loss of any solution/spray/acid from flask
or
to allow gas to escape
or
spurting
or
to stop any <u>solids/liquids</u> getting in/out

(b) (i) 0.017
(ii) answer between 0.37 and 0.4

3. (a) E_h = cmΔT
= 4.18 × 0.5 × 82
= ±171 kJ

number of moles required = $\frac{171}{1367}$

answer 0.12 or 0.125 or 0.13 moles

(b) *Any two from:*
heat lost to surroundings
incomplete combustion (of alcohol)
ethanol impure
loss (of ethanol) through evaporation

4. (a) $^{89}Sr \longrightarrow {}^{89}Y + \beta$

or

$^{89}_{38}Sr \longrightarrow {}^{89}_{39}Y + {}^{0}_{-1}e$

(b) (i) no effect/no change
(ii) $\frac{89}{160} \times 10 = 5.56g$ or 5.6g

(c) ¼ **or** 0.25 **or** 25%

5. (a) (24 litres) 24,000 cm^3 → 6·02 × 10^{23}

(0·110 litres) 110 cm^3 → 110/24000 × 6·02 × 10^{23}
= 2.76 × 10^{21}

(b) $CH_3CH_2OH + O_2 \rightarrow CH_3COOH + H_2O$

or

$CH_3CH_2OH + O_2 + 4H^+ + H_2O \rightarrow 2H_2O + CH_3COOH + 4H^+$

(c) catalyst/reactants different state

6. (a)

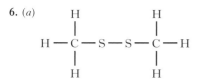

or any structure for an expansion of the shortened structural formula $CH_3S_2CH_3$ containing
- 6 hydrogen atoms, valency 1
- 2 carbon atoms, valency 4
- 2 sulphur atoms, valency 2 or 4 or 6

(b) (i) *Either:*

moles Cl_2 $0·010 × 0·0294 = 2·94 × 10^{-4}$

moles H_2S $2·94 × 10^{-4}/4 = 7·35 × 10^{-5}$

conc^n H_2S $\dfrac{7·35 × 10^{-5}}{0.05}$

$= \underline{1·47 × 10^{-3}}$

or

candidates may use a "titration" formula of which an example is shown below.

$\dfrac{c_1 v_1}{b_1} = \dfrac{c_2 v_2}{b_2}$

(ii) *This question is divided into two separate marks, the first subdivided:*

First Mark

permanent dipole-permanent dipole attractions or polar-polar attractions/forces ½

weak intermolecular bonds/forces ½

Second Mark

If pd-pd named then:
mention of difference in electronegativities or indication of polar bonds or indication of permanent dipole

If VdW/LDF named:
instantaneous dipoles or temporary dipoles or uneven distribution of electrons or electron wobbles

7. (a) w=9 x=6 y=2 z=2
or
$C_9H_6N_2O_2$

(b) no elimination of a small molecule (such as water)
or the monomers have <u>added across</u> the (N=C) double bond
or only one product molecule formed
or joined across the (N=C) double bond

(c) dotted lines between H/N or H/O on adjacent polymer chains

8. (a) amide link or peptide link or peptide bond

(b)

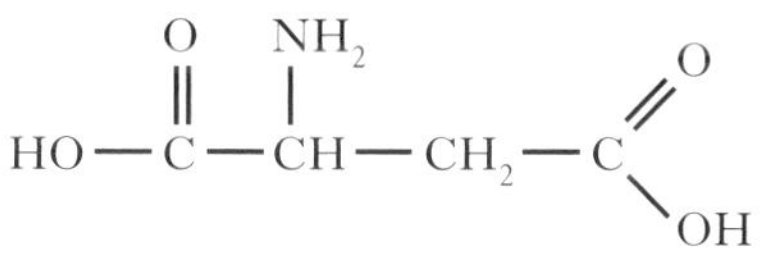

$$HO - \overset{\overset{\displaystyle O}{\|}}{C} - \underset{}{\overset{\overset{\displaystyle NH_2}{|}}{CH}} - CH_2 - C\overset{\displaystyle O}{\underset{OH}{\diagup}}$$

(c) essential

(d) 1 mark for wet paper towel (condenser) or cold finger test tube
1 mark for use a condenser
1 mark for raise the test-tube so that a greater length of the test-tube is above the hot water, but with the reaction mix still immersed or lower the level of the water

9. (a) ½ mark for bromine (water)/iodine (solution)
½ mark for Oleic decolourises
or stearic does not decolourise/decolourises slowly

(b) octadec −9, 12, 15 −trienoic acid
octadeca −9, 12, 15 −trienoic acid

(c) **either** O^-Na^+ **or** CO^-Na^+ **or** COO^-Na^+ **or** O^- **or** $C-O^-$ **or** COO^-

10. (a) air

(b) methyl methanoate

(c) **either**

$HCOOH \quad \rightarrow \quad HCONH_2$
1 mole   1 mole
46g   45g
1·38g   <u>1·35g</u>

% yield $= \dfrac{0.945g}{1.35g} × 100$

$= 70\%$

or

moles $HCOOH \rightarrow 1·38/46 = 0·03$
moles $HCONH_2 \rightarrow 0·945/45 = 0·021$

$HCOOH \quad \rightarrow \quad HCONH_2$
0·03 moles $\rightarrow$ 0·03 moles

% yield $= 0·021/0·03 × 100$

$= 70\%$

11. (a) (i) 3-methyl butan-2-ol

(ii)

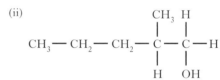

(b) (i) $4BF_3 + 3 NaBH_4 \rightarrow 2B_2H_6 + 3 NaBF_4$

(ii) -36 kJ
-1274 kJ
$3 × -286 = -858$ kJ

½ mark for each correct enthalpy change

½ mark for addition of 3 sensible numbers

-2168 kJ mol^{-1}

(c) 143444 **or** −143444 **or** 143000 **or** −143000 **or** 145000 **or** −145000
or
143 MJ

12. (a) to allow the potato discs/catalase to reach the pH of the buffer
or
to allow buffer to soak/diffuse into the potato disc
or
to allow the enzyme/potato to reach the same pH as the surrounding solution
or
to allow the enzyme/potato to acclimatise

(b) hydrogen peroxide/H_2O_2

(c) the enzyme is denatured
or
the enzyme changes its shape
or
enzymes work best at an optimum pH
or
too acidic for enzyme to function

13. (a) $Q = I \times t = 5.0 \times 60 \times 32 = 9600$ C

    1 mol F_2 needs 2 moles of electrons = <u>2 × 96 500 C</u>

    193000 C → 38g
    9600 C → 1.89g

(b) (i) exothermic or heat given out or ΔH is −ve or ΔH<0

    (ii) graph shows as pressure increases/concⁿ C_2F_4
    decreases.

       line sloping <u>downward</u>

(c) depletion/break down of the ozone layer

14. (a) (i) $[H^+(aq)] = 1 \times 10^{-5}$ mol l⁻¹

    (ii) *The marks for this question are divided into two separate marks:*

       The first mark is awarded for the ammonia/ammonium equilibrium:

       ½ mark for
       $NH_3 (aq) + H_2O (\ell) \rightleftharpoons NH_4^+ (aq) + OH^-(aq)$

       1 mark if it has been shown that the position of this equilibrium is such that the ammonium ions tend to remove OH^- ions from solution
       e.g. $NH_4^+ (aq)\ OH^- (aq) \rightarrow NH_3 + H_2O$
       or suitable description in words

       The second mark is awarded for the water equilibrium:
       ½ mark for
       $H_2O (\ell) \rightleftharpoons H^+ (aq) + OH^- (aq)$

       1 mark if it has been shown that water molecules dissociate resulting in an increased H^+ ion concentration
       e.g. $H_2O (\ell) \rightarrow H^+ (aq) + OH^- (aq)$
       or suitable description in words

(b) answers showing an appreciation that a <u>large volume</u> or large number of moles of gas is produced
    or
    there is an increase in the number of moles of gas
    or
    oxygen gas is produced which can support combustion
    or
    it is an oxidising agent

15. (a) (i) to keep the current constant or to adjust the current

    (ii) the current and the time

(b) (i) recycle/reuse the <u>SO_2</u> and/or <u>H_2O</u>
       or O_2 can be sold

    (ii) $H_2O \rightarrow H_2 + \frac{1}{2}O_2$ **or** $2H_2O \rightarrow 2H_2 + O_2$

16. (a)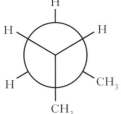

(b) 2-methylbutane
    or
    methylbutane

CHEMISTRY HIGHER 2013

SECTION A

1. A	11. B	21. D	31. C
2. C	12. C	22. B	32. A
3. B	13. D	23. D	33. A
4. B	14. C	24. B	34. C
5. D	15. D	25. C	35. D
6. A	16. D	26. A	36. C
7. D	17. A	27. D	37. D
8. C	18. A	28. B	38. B
9. B	19. C	29. A	39. B
10. A	20. B	30. A	40. C

SECTION B

1. (a) reforming / reformation

(b) 2,2,4-trimethylpentane

(c) it has branches
    or shorter chain

(d) methanol toxic (**or** poisonous **or** makes you blind)
    or CO_2 / CO emissions
    or (burns to) produce greenhouse gases
    or methanol is corrosive
    or methanol releases less energy than petrol

2. (a) purple
    or (pink) to colourless
    or purple (pink) decolourises

(b) (i) 58 (°C)

    (ii) (colour) change too gradual

(c) more molecules (particles) have <u>enough energy to collide successfully</u>
    or
    more molecules with energy greater than the activation energy

3. (a) $3Al + 3NH_4ClO_4 \rightarrow Al_2O_3 + AlCl_3 + 3NO + 6H_2O$

(b) 0·255 (g) or 0·26 (g)

4. (a) (i) $K(g) \rightarrow K^+(g) + e^-$
    (ii) answers can be given either in terms of potassium or of chlorine
       K has more shells/levels or electron further from nucleus
       this leads to greater shielding/screening
       so less energy required to remove electron/weaker attraction for the electron
       or
       Cl has fewer shells or electron closer to nucleus
       this leads to less shielding/screening
       so more energy required to remove electron/stronger attraction for the electron

(b) 8

5. (*a*) any reasonable molecular structure consisting of two carbon and two nitrogen atoms in which each carbon forms four bonds and each nitrogen three bonds.

For example

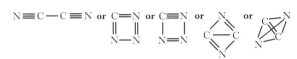

(*b*) method for removal of carbon dioxide which would work (eg by bubbling through lime-water or any alkaline solution or passing over soda lime or lithium oxide etc.)

method for collection and measurement of gas which would work (eg in gas syringe or in an upturned measuring cylinder over water)

6. (*a*) trichloromethane is polar **or** tetrachloromethane is non-polar

trichloromethane is capable of forming (permanent) dipole/(permanent) dipole attractions

tetrachloromethane is only capable for forming van der Waals'/London Dispersion Forces

water is polar solvent/like dissolves like/is a good solvent for polar substances

(*b*) absorbs (harmful) UV
or reduces (or stops) UV reaching earth or protects (us) from UV
or filters the UV

7. (*a*) Tollen's or acidified dichromate or Fehling's or Benedict's

(*b*) (i) 8
(ii) oxidation

8. (*a*)

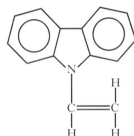

(*b*) in photocopiers
or in laser printers
or as a photoconductive (material)

9. (*a*) ethane-1,2-diol contains two/more −OH groups
or
ethane-1,2-diol forms more/stronger hydrogen bonds
or
more energy needed to break the hydrogen bonds

(*b*) 2-methylbut-2-ene or methylbut-2-ene

(*c*) benzene-1,3-dicarboxylic (acid)

10. (*a*) (i) amino group or amine group
(ii)

$$HO - CH_2 - CH_2 - N \begin{array}{c} CH_2-CH_3 \\ \\ CH_2-CH_3 \end{array}$$

(*b*) 25 (minutes)
or
8·0 to 8·4 (minutes)

11. (*a*) esterification or condensation

(*b*)
1 mol salicylic acid	→	1 mol aspirin	
138 g	→	180 g	
5·02 g	→	$\dfrac{180}{138}$	× 5·02

$\Rightarrow$ theoretical yield = 6·55 g

$\Rightarrow$ % yield = $\dfrac{2·62}{6·55}$ × 100

= 40%

or

1 mol salicylic acid → 1 mol aspirin

number of moles of salicylic acid = 5·02/138 = 0·0364

number of moles of aspirin = 2·62/180 = 0·0146

% yield = 0·0146/0·0364 × 100

= 40 %

(*c*) (it is a salt of a) strong base and a weak acid
or a valid explanation of the equilibria involved

12. (*a*)
Mass number	0
Atomic number	1

(*b*) $2·52 \times 10^9$ years

13. (*a*) (i) partially dissociated
or
not completely ionised

(ii) equilibrium would shift to right/forward
or
more products formed

(*b*) 0·29 g

Method one

moles H_2S = $\dfrac{0·079}{24}$ **or** 0·00329

moles FeS = 0·00329
GFM FeS = 87·9 g
mass FeS = 87·9 × 0·00329
mass FeS = 0·29 g

Method two

$$\dfrac{Vol\ H_2S}{24} \leftrightarrow \dfrac{Mass\ FeS}{87·9}$$

$$1 \leftrightarrow \dfrac{87·9}{24}$$

$$0·079 \leftrightarrow \dfrac{87·9}{24} \times 0·079$$

$$0·079 \leftrightarrow 0·29\ g$$

14. (*a*) (i) synthesis gas
or
syngas

(ii) (+)206 (kJ mol⁻¹)

(*b*)
temperature	decrease/keep the same/increase
pressure	decrease/keep the same/increase

15. (a) mcΔT   = $0.050 \times 4.18 \times 4.5$
             = -0.94 kJ
    moles H_2O = 0.025
    0.025 mol $\leftrightarrow -0.94$ kJ

    1 mol $\leftrightarrow \dfrac{-0.94}{0.025}$
             = -38 kJ mol^{-1}

(b) lid added/use polystyrene (plastic) cup/
    insulate beaker/closed container heatproof container

(c) initial temperature of (both) solutions or the average start
    temperature

    maximum/final/end temperature (of mixture)

16. (a) $I_2 + 2e^- \rightarrow 2I^-$

(b) (i) first titre is a rough (or approximate) result/practice
        or
        first titre is not accurate/not reliable/rogue
        or
        first titre is too far away from the others
        or
        you take average of concordant/close results

    (ii) 0.045 or 0.05 (mol l^{-1}) if working correct

    either

    moles S_2O_3 $0.10 \times 0.01815 = 0.001815$

    moles of I_2 $\dfrac{0.001815}{2} = 0.0009075$

    concentration $I_2 = \dfrac{0.0009075}{0.0200}$

    concentration $I_2 = 0.045$

17. (a) Q = It

    Q = $0.5 \times 2 \times 60 \times 60$

    Q = 3600C

Charge		Mass of Pb
2×96500C	$\rightarrow$	207.2 g
3600C	$\rightarrow$	3.86 g

(b) $PbO_2(s) + SO_4^{2-}(aq) + \mathbf{4H^+ + 2e^-} \rightarrow PbSO_4(s) + \mathbf{2H_2O}$

18. (a)

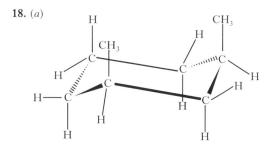

(b) (i) the bigger the group the greater the strain
        or
        the larger the (halogen) atom the greater the strain
        or
        the more atoms in a group, the greater the strain
        or
        any other statement which is consistent with the
        values presented

    (ii) 7.6 (kJ mol^{-1})

CHEMISTRY HIGHER 2014

SECTION A

1. A	11. C	21. B	31. C
2. D	12. D	22. D	32. B
3. D	13. B	23. C	33. A
4. B	14. A	24. D	34. C
5. C	15. B	25. B	35. D
6. A	16. C	26. D	36. B
7. C	17. C	27. D	37. A
8. C	18. B	28. A	38. B
9. A	19. B	29. A	39. D
10. A	20. A	30. B	40. D

SECTION B

1. (a) Completed table in order:
    Metallic (metal)
    Network (lattice)
    Covalent
    Molecular (discrete)

(b) Delocalised / free electrons

(c) Increasing nuclear charge/increasing number of protons
    (pulls electrons closer)

2. (a) Naphtha

(b) (i)

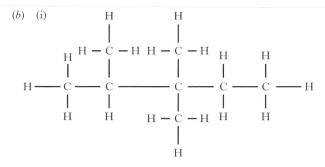

    (ii) Aromatic (hydrocarbons) **or** cycloalkanes or cyclic
         (hydrocarbons)

(c) (i) Two factors are:
        Not too high as to denature enzyme
        High enough to give fast reaction optimum/most
        efficient temperature

    (ii) Oxygen to hydrogen ratio has decreased
         or hydrogen to oxygen ratio has increased
         or hydrogen has been gained

    (iii) **Either**

Glucose	**Ethanol**
1 mole	2 moles
180 g	92 g
	$\dfrac{92 \times 1000}{180}$
1000 g	$= 511.11$ g

    % Yield = $\dfrac{445 \times 100}{511.11}$

           = $87(.1)$ %

    or

    Moles glucose = $\dfrac{1000}{180} = 5.56$

    Actual Yield

    Moles ethanol = $\dfrac{445}{46} = 9.67$

Theoretical Yield

Glucose   **Ethanol**

5·56 mole   11·12 moles

% Yield $= \dfrac{9·67}{11·12}$

$= 87\%$

3. (a) (i) Example e.g. 20 cm³ KI solution plus 5 cm³ water to explain dilution with water keeping (total) volume constant.

   (ii) *Any two from:*
   - Start timing as hydrogen peroxide is added
   - Use more accurate measuring equipment such as syringes pipettes, burettes, smaller measuring cylinder) to measure the solutions
   - Use a white tile under beaker
   - Stirring/swirling
   - Repeat the experiment

   (b) Collision must occur with sufficient energy/force to break bonds (Answer must show understanding of activation energy)
   or
   Collision must occur with suitable geometry

4. (a) H-F has hydrogen bonds and F-F has van der Waals'/London dispersion forces

   Hydrogen bonds stronger

   Hydrogen bonds caused by:
   (large) difference in electronegativity
   or indication of polar bonds
   or indication of permanent dipole

   van der Waals' forces caused by:
   Temporary dipoles
   or uneven distribution of electrons
   or electron cloud wobble/Movement of electrons

   (b) Any pH greater than 7
   If range given it must not include 7

5. (a) (i) Condensation

   (ii) Any answer that indicates that ethanoic acid has only one functional group (so the chain cannot continue)
   or
   A monomer must contain 2 functional groups
   or
   It is not a diacid

   (b) Conducts (electricity)

6. OH⁻ ions react with H⁺ ions
   H⁺ concentration decreases
   Equilibrium shifts to right
   More $C_{14}H_{14}N_3SO_3^-$ **or** less $C_{14}H_{15}N_3SO_3$

   Then becomes (more) yellow or less red (must be linked to a valid reason from above.)

7. (a) ^1_1H or ^1_1p

   (b) Proton is produced
   or neutron splits to give proton (and electron)
   or nucleus contains one more proton

   (c) (i) From graph, half-lives = $4·3 \pm 0·1$

   Age = $4·3 \times 5700 = 24510$
   (23940 – 25080)

   (ii) Radioactivity/Amount of C-14 too low.
   Too short a half-life; Too little C-14 remains.

8. (a) $E_h = cm\Delta T$
   Correct substitution of data
   $= 4·18 \times 0·21 \times 50$
   $= \pm 43·89$ kJ (no units required)

   or

   $= 4·18 \times 210 \times 50$
   $= 43890$ J (no units required)

   Then

   65 kJ → 56 g
   43·89 kJ (44) → $\dfrac{56}{65} \times 43·89$ (44)

   $= 37·81$ g (38 g)

   or

   Moles required $= \dfrac{43·89}{65} = 0·67$

   Mass $= 0·67 \times 56$
   $= 37·52$ g

   (b) $Ca(s) + \frac{1}{2}O_2(g) \rightarrow CaO(s)$ (reversed)
   $\Delta H = +635$ kJ mol⁻¹

   $H_2(g) + \frac{1}{2}O2(g) \rightarrow H_2O(\ell)$ (reversed)
   $\Delta H = +286$ kJ mol⁻¹

   $Ca(s) + O_2(g) + H_2(g) \rightarrow Ca(OH)_2(s)$
   $\Delta H = -986$ kJ mol⁻¹

   $Ca(OH)_2(s) \rightarrow Ca(OH)_2(aq)$
   $\Delta H = -82$ kJ mol⁻¹

   Add together $= -147$ (kJ mol⁻¹)

9. (a) (i) Concentrated sulphuric acid

   (ii) Reaction mixture and/or ester produced is flammable
   Any mention of flammable/burning

   (iii)

   H—C—O—C—C—H (structure with O double bonded to first C, and H's on carbons)

   or
   $HCOOCH_2CH_3$
   or
   partially shortened structural formula

   (b) $CHCl_3 + 4NaOH \rightarrow HCOONa + 3NaCl + 2H_2O$
   or multiples including 1/2

10. (a) Heterogeneous

    (b) $CH_4 + 2H_2O \rightarrow CO_2 + 4H_2$

    (c) (i) $Q = It = 200 \times 60 \times 30$
    $= 360,000$ C

    1 mol H_2 needs 2 moles electrons
    $= 2 \times 96500 = 193,000$ C

193,000	24 litres
360,000	$\dfrac{360,000}{193,000} \times 24$

    $= 44·77$ litres

    (ii) Doesn't produce CO_2

    or CO_2 is bad for the environment
    or no polluting by-product

    or no by-product to separate
    or getting pure hydrogen
    O_2 produced by electrolysis
    If global warming given must be linked to CO_2

11. (*a*) w = 10, x = 5, y = 2, z = 1

  (*b*) 4-methylpentan-2-one

  (*c*) (CFCs) destroy/deplete/damage ozone (layer)/ makes holes in ozone layer

  (*d*) (i) Hydrogenation

    (ii)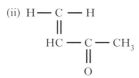

    Or other correct drawing of this structure

    Accept the final product of the reaction

    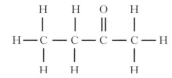

12. (*a*) Fibrous

  (*b*) Peptide link correctly identified including just

  (*c*) Hydroxyl

  (*d*) (i) Glycerol

    or propan(e)−1,2,3-triol
    or glycerin(e)

    (ii) From a hydrogen connected to an oxygen or nitrogen to another oxygen (includes the carbonyl oxygen) or nitrogen
    Hydrogen bond correctly drawn

13. (*a*) Top of meniscus /curve was read
    Use a bright light behind burette

  (*b*) Blue/black or to purple colour

  (*c*) It was the rough titre
    or wasn't done accurately

  (*d*) **Either**

    moles I_2 = moles Vit C
    = 0·00125 × 0·0254 = 0·0000317 (0·00003)(0·000032)

    Scale up to 1 litre:
    0·0000317 × 50 = 0·00159 (0·0015)(0·0016)

    Calculation of mass = moles × 176

    = 0·00159 × 176 = 0·279 g (0·264)(0·282)

    or

    Candidates may use a "titration" formula of which an example is shown below

    $$\frac{C1 \times V1}{b1} = \frac{C2 \times V2}{b2}$$

    $$\frac{C1 \times 20}{1} = \frac{0·00125 \times 25·4}{1}$$

    Rearrangement:

    $$C1 = \frac{0·00125 \times 25·4}{20} = 0·00159$$

    Calculation of mass = moles × 176

    = 0·00159 × 176 = 0·279 g

14. (*a*) 222 g gives 9×24

    1 g gives $\frac{9 \times 24}{222}$ = 0·973 litres

    Or 973 cm^3

  (*b*) $2CO + 3CO_2 + 4H_2O + 2N_2$

15. (*a*)

    O⁻—C—C——C—OH with O double bond, NH₃⁺, H

  (*b*) OH⁻ concentration = 1 × 10⁻⁵ mol l⁻¹
    H⁺ concentration = 1 × 10⁻⁹ mol l⁻¹
    pH = 9
    Positive zwitterion forms
    Moves to negative electrode